FARMERS OF THE WORLD

FARMERS OF THE WORLD

THE DEVELOPMENT OF AGRICULTURAL EXTENSION

Edited by

EDMUND deS. BRUNNER

IRWIN T. SANDERS, *and*

DOUGLAS ENSMINGER

Essay Index Reprint Series

BOOKS FOR LIBRARIES PRESS

FREEPORT, NEW YORK

INTERNATIONAL STANDARD BOOK NUMBER:
0-8369-2182-8

LIBRARY OF CONGRESS CATALOG CARD NUMBER:
73-134062

PRINTED IN THE UNITED STATES OF AMERICA

FOREWORD

THE PRIMARY AIM of this book is to discuss the most effective general approach which a government or a private agency can use in helping rural people solve their everyday problems. This approach is commonly called Extension. Considerable space is devoted to *specific* ways and means of carrying on extension but these are incidental to the primary aim. The basis of discussion throughout is within the democratic framework; namely, that the end of all extension activity is the growth and improvement of the individual rather than the fashioning of automatons who merely obey orders handed down from above.

The authors of the chapters which follow have applied to the areas which they have described the inquisitive approach of the social scientist. They have asked two questions: First, what is the cultural background against which extension work must be carried on? Secondly, how can innovations desired by the people most easily become a part of their way of life? Every author has used his own approach because in so doing he can best tell the story of extension in the part of the world about which he is writing. (The opinions expressed are those of the individual and are not necessarily those of the agency which he represents.)

The reader will be surprised at the similarity in the experience of the writers, whether in the Balkans, in the Middle East, or in India. These men have arrived at their conclusions independently. And, as the last chapter draws together the several strands that run throughout the whole book, the reader can see for himself how essential it is that extension systems be thought of primarily as servants of the people and that the spreading of science does not necessarily mean from laboratory to farm but can often mean from the farm of a successful cultivator to another farm. The reader will also see that any extension developed out of line with the basic institutions and attitudes of an area is usually fruitless, as it means imposing ideas or patterns upon people unready and unwilling to accept them. There is no extension unless people are changed, and there is little constructive change unless the people cooperate. Extension moves forward only as it is charged with the dynamic energy of earnest men and women seeking answers to problems of everyday life.

THE EDITORS

CONTENTS

ABOUT THE AUTHORS   ix

INTRODUCTION

1. WHAT EXTENSION IS, *by Douglas Ensminger and Irwin T. Sanders*   1

NONLITERATE SOCIETIES

2. DIVERSITY AND CHANGE IN THE CULTURE OF NONLITERATE PEOPLES, *by Solon T. Kimball*   8

3. EXTENSION WORK IN THE PACIFIC ISLANDS, *by Felix M. Keesing*   19

PEASANT SOCIETIES

4. CHARACTERISTICS OF PEASANT SOCIETIES, *by Irwin T. Sanders*   37

5. PROMOTING COOPERATIVE AGRICULTURAL EXTENSION SERVICE IN CHINA, *by Hsin-Pao Yang*   46

6. EXTENSION EXPERIENCE IN INDIA, *by D. Spencer Hatch*   61

7. EXTENSION WORK AMONG THE ARAB FELLAHIN, *by Afif I. Tannous*   78

8. EXTENSION IN THE BALKANS, *by Clayton E. Whipple*   101

9. EXTENSION WORK IN LATIN AMERICA, *by Charles P. Loomis*   117

EURO-AMERICAN SOCIETY

10. EURO-AMERICAN RURAL SOCIETY, *by Carle C. Zimmerman*   138

11. EXTENSION IN THE UNITED KINGDOM, *by Robert Rae*   153

12. AGRICULTURAL EXTENSION SERVICES IN NORTH-WEST EUROPE, *by P. Lamartine Yates and L. A. H. Pieters*   165

Contents

13. AGRICULTURAL EXTENSION IN THE UNITED
STATES, *by Edmund deS. Brunner and C. B. Smith*   180

CONCLUSION

14. THE ROLE OF EXTENSION IN WORLD RECONSTRUC-
TION, *by M. L. Wilson and Edmund deS. Brunner*   193

Index   201

ABOUT THE AUTHORS

EDMUND DES. BRUNNER, Ph.D., L.H.D., has been Professor at Columbia University since 1931, collaborator in the Bureau of Agricultural Economics of the U.S. Department of Agriculture since 1936 and Adviser in the Department's Extension Service since 1942. He directed town and country studies for the Institute of Social Research (Rockefeller) from 1920 to 1933, and in such capacity conducted eight nation-wide rural socioeconomic studies in the United States and one in Korea. He has made other field trips and traveled in Japan, China, the Philippines, India, Egypt, New Zealand, Australia, and several Pacific islands. He is the author of *American Agricultural Villages, Rural Korea, Rural Social Trends* (with J. H. Kolb), *Rural Trends in Depression Years* (with Irving Lorge), *Immigrant Farmers and Their Children*, and twenty other volumes.

DOUGLAS ENSMINGER, Ph.D., has been Principal Social Scientist, Bureau of Agricultural Economics, United States Department of Agriculture since 1939 and in charge of Rural Sociology Extension of the Department's Federal Extension Service since 1942, where among other responsibilities he is charged with the supervision of research work in community organization. Formerly he served with the Farm Security Administration and was Assistant State Supervisor of Research at the University of Missouri. He has published a number of community and institution studies and is the author of *The Rural Church in Missouri, Rich Land, Poor People, Diagnosing Rural Community Organization.*

D. SPENCER HATCH, Ph.D., is with the World Service Department of the Young Men's Christian Association in Mexico as an Agricultural Extension missionary. Formerly he served in a similar capacity in India for many years. He has published, among other works, *Up from Poverty in Rural India* and *Further Upward in Rural India.*

FELIX M. KEESING, D.Litt., is Professor of Anthropology at Stanford University, California. Formerly he was successively from 1934 to 1943 Assistant and Associate Professor of Anthropology at Hawaii University and Professor and Chairman of the Department of Anthropology and Sociology at the same University. Among his specific activities have been studies of the culture of the Maori (the native people of New Zealand), and of acculturation in New Zealand, in the Pacific, and of the Menomini Indians in the United States.

His publications include *The Menomini Indians of Wisconsin, The South Seas in the Modern World*, and *Hawaian Homesteading on Malokai.*

SOLON T. KIMBALL, Ph.D., is Head of the section of Community Organization and Activities of the War Relocation Authority. He has served (1936–1942) with the U.S. Soil Conservation Service and with the Indian Service as Social Economist. He was in charge of the socioeconomic surveys for the Navajo Reservation. He was (1934–1936) research assistant in the Department of Anthropology at Harvard University, prior to which he held a fellowship for study in the Irish Free State. He collaborated with C. M. Arensberg in *Family and Community in Ireland.*

CHARLES P. LOOMIS, Ph.D., is Head of the Department of Sociology and Anthropology at Michigan State College. Prior to that he was Head of the Division of Extension and Training in the Office of Foreign Agricultural Relations. He grew up in the Mexican-American section of New Mexico and in addition to European experience worked as an extension specialist in Central and South America for the United States Department of Agriculture. He has held the following positions related to his contribution in this book: Teacher of Agriculture at Las Cruces, New Mexico; Assistant Professor of Sociology, North Carolina State College; Leader of the Section on Studies of the Community and Social Organization. Division of Farm Population and Rural Welfare Bureau of Agricultural Economics, U.S. Department of Agriculture; and Visiting Lecturer in Sociology at Harvard University.

His works include *Fundamental Concepts of Sociology* (a translation); *Culture of a Contemporary Rural Community—El Cerrito, New Mexico;* and articles on "Extension Work at Tingo Maria" and "Spanish Americans: the New Mexican Experiment in Village Rehabilitation."

L. A. H. PIETERS, is Agricultural Attaché of the Netherlands Embassy in Washington. He graduated as an Agricultural Engineer from Wageningen University, Holland, and has served as Secretary to the Netherlands Minister of Agriculture and Fisheries, The Hague, as well as Agricultural Attaché in the Netherlands Legation in Brussels.

ROBERT RAE, B.Sc.Agric., is Agricultural Attaché at the British Embassy in Washington, D.C. He was educated at Edinburgh University and the East of Scotland College of Agriculture. Since 1933 he has been Professor of Agriculture and director of the University farm at the University of Reading, which has the largest teaching Argicultural Department of any British University. Prior to 1933 he was Professor of Crop and

Animal Husbandry at the Queen's University of Belfast, Northern Ireland, head of the Northern Ireland Ministry of Agriculture's Crop and Animal Husbandry Research Division, and the first director of the Agricultural Research Institute, Hillsborough, County Down. During the present war Professor Rae has acted as one of the technical advisers of the British Ministry of Agriculture and has also been a member of a number of the Committees of the Berkshire War Agricultural Executive Committee. He is a member of many educational bodies and farmers' societies. He has had many years of practical experience of farming in Scotland, Ireland and England.

He has contributed frequently to British agricultural journals the results of research and experimental work and other articles of general agricultural interest. He has traveled widely in America on several occasions.

IRWIN T. SANDERS, Ph.D., is professor and Acting Head of the Sociology Department, University of Kentucky. He was for six years connected with The American College of Sofia, Bulgaria (serving as Dean from 1934 to 1937), where he did extensive research in the socioeconomic phases of peasant life. He has held temporary appointments with the Office of Foreign Agricultural Relations and with the Extension Service of the U.S. Department of Agriculture. His sociological writings include a chapter on "Community and Education" in *Sociological Foundations of Education;* a monograph—*Alabama Rural Communities;* and a number of articles for professional journals in the Balkans and the United States.

C. B. SMITH, M.S. and D.S., served from 1914 to 1938 with the Extension Service of the Department of Agriculture, successively as Chief of Office of Extension Work, North and West; Chief of the Office of Cooperative Extension Work; and Assistant Director of Extension Service. Previously he was Horticultural Editor in the Office of Experiment Stations for the Department of Agriculture and for six years in charge of the Section of Field Studies and Demonstrations in the Office of Farm Management of the Department.

His works include *Farmers' Cyclopedia of Agriculture, Farmers' Cyclopedia of Livestock* (both with E. V. Wilcox), *The Agricultural Extension System of the United States* (with M. C. Wilson), and many farmers' bulletins, circulars and Extension reports.

AFIF I. TANNOUS, Ph.D., is a native of Syria, who, after graduating from the American University of Beirut, had a wide extension experience in the Anglo-Egyptian Sudan with the British government. He conducted investigations in Greece, had charge of rural field work for the Near East Foundation in Palestine, and directed the Village Welfare Service of the American University of Beirut. He completed his graduate work in Rural Sociology at St. Lawrence and Cornell Universities, and is on leave from the University of Minnesota. He is connected with the Office of Foreign Agricultural Relations where his work covers the socioeconomic aspects of rural life in the Middle East. He has written extensively on the village problems of the Middle East, his latest contribution being "The Arab Village Community of the Middle East" in *The Smithsonian Institution Report for 1943*.

CLAYTON E. WHIPPLE, M.S.Ed., is in charge of the Balkan and Near Eastern Section in the Office of Foreign Agricultural Relations, Department of Agriculture. From 1929 to 1940 he was Director of Rural Education of the Near East Foundation in the Balkans and the Near East. He served (1934–1940) as adviser on extension and rural education to the Bulgarian government and was consultant in the same fields to other Balkan governments. He was a member of the Secretariat at the United Nations Conference on Food and Agriculture at Hot Springs, Virginia, 1943. He has written articles on Balkan agriculture and extension for numerous professional journals, many of them in the Balkans.

MILBURN LINCOLN WILSON has been Director of Cooperative Extension Work, United States Department of Agriculture, since February 1, 1940. His career includes that of tenant farmer; County Agent in Montana; State County Agent Extension Leader in the same state; Head of Division of Farm Management and Costs, United States Department of Agriculture; Professor at Montana State College, with an interim as Technical Consultant in Russia on introducing methods of handling wheat on a large scale in the North Caucasus. Mr. Wilson contributed a great deal toward blueprinting agricultural reforms of the early 1930's, was successively Wheat Production Adjustment Administrator for the AAA, Director of Subsistence Homestead Division, Assistant Secretary and later Under Secretary of Agriculture, dealing continuously in these capacities with relationships between the Federal Department of Agriculture and the states. He has guided the educational phases of the wartime national nutrition program.

   Publications: *Farm Relief and the Domestic Allotment Plan, Democracy Has Roots*, "Nutrition, Food Attitudes and Food Supply," an article in *Post-war Economic Problems*, and "Thomas Jefferson—Farmer," in *Proceedings* of the American Philosophical Society.

PAUL LAMARTINE YATES is an Englishman at present working in Washington D.C. on loan with the United Nations Interim Commission on Food and Agriculture. During school years he traveled widely in Europe and South Africa. At Cambridge University he took the Historical and Economics triposes and afterwards went to Europe studying chiefly in Berlin but also Geneva and elsewhere. Returning to England, he became secretary to a body known as "The Astor-Roundtree Agricultural Group" which after three years of research produced a most comprehensive volume on *British Agriculture*. Then, at the Group's request, he made an extensive tour of the farming regions of Western Europe, visiting and talking with several hundred farm families in six different countries. In 1940 he joined the British Ministry of Economic Warfare to sift intelligence on the food and agriculture situation in Nazi-dominated Europe. Later, the British Ministry of Agriculture secured his part-time service.

His publications include *Food Production in Western Europe, Food and Farming in Post-War Europe* (with D. Warriner), and *Commodity Control* (in collaboration with a Fabian Society group).

HSIN-PAO YANG, D.Ed., is a Research Fellow of the Institute of Adult Education at Teachers College, Columbia University. Formerly he was a trainee in Agricultural Extension of the United States Department of Agriculture. Prior to that he was Director of Mass Education at Fukien Province, China, and assisted in university extension work largely in rural areas conducted by Fukien Christian University, Foochow. He also served as Educational Director of Civilian Defence in Fukien Province. He has done further graduate work at Graduate College, Iowa State College; the University of Chicago; and the Graduate School of the United States Department of Agriculture.

He has several manuscripts awaiting publication including an outline for community analysis and institutional study for the Chinese community and a Study of Cooperative Extension System of the United States.

CARLE C. ZIMMERMAN, Ph.D., is Professor in Rural Sociology at Harvard University and formerly held the same title at the University of Minnesota. He has served with the United States Army as a major in the Air Corps. He has written many articles and experiment station bulletins on sociology and his books include *Source Book in Rural Sociology* (with P. Sorokin and C. J. Galpin), *Consumption and Standards of Living*, and *Changing Community*.

Chapter 1 · WHAT EXTENSION IS · By Douglas Ensminger and Irwin T. Sanders

THE TASK OF EXTENSION WORK is to help rural families apply science to the day-by-day routine of farming, homemaking, and other aspects of rural living. Extension is away-from-the-class-room education.

In the great majority of the countries around the world special agencies, public and private, communicate with, and carry on educational activities among, farm people. The type of organization varies from the highly systematized and coordinated governmental plan found in the United States to the limited extension programs of the Middle East. The differences in form of organization and methods employed by the various agencies in conducting the work are due in part to variations in land-tenure patterns, credit facilities, cultural background, and the nature of the higher educational institutions offering courses in agriculture and related fields.

Extension education employs the principle of cultural variation and culture change, abundantly illustrated in the chapters which follow. It also emphasizes working *with* the people rather than *for* them and selects for treatment those problems which the people themselves recognize as important. At times, it goes a step further and teaches people to recognize as problems for solution conditions which they had accepted as inevitable or about which they had previously felt little concern. Thus extension education teaches people *what* to want as well as *how* to work out ways of satisfying these wants.

In England, prior to the war, extension was primarily a local matter, with little centralization on a national scale. Extension work in the United States is cooperative. The Federal Department of Agriculture, the state agricultural colleges, the county government and the people themselves are the cooperating groups. In some other countries, such as Italy, extension was dominated from the top down, farmers being told what they had to do in the interest of the Fascist regime. In Belgium the program of rural improvement had a religious as well as a national basis, making use of church cooperatives to a great extent. In India and China, rural reconstruction was as much the affair of private agencies as it was the concern of the government. There religious and educational institutions have spasmodically explored the possibilities, and their experiences now show the shape which improved national extension work in these countries should take.

The "result demonstration" is a basic cornerstone in extension teaching, much of which has been by use of object lessons. When an improved variety of wheat is developed, the extension personnel works closely with the farmers in determining whether it is adaptable to local farming conditions. In one field the farmer may plant the new variety; next to it he plants the old variety. The farmer and his neighbors themselves judge if the new variety is superior. If it is, they will purchase the new variety seed and plant it the next year. This knowledge and spirit passes from enthusiast to enthusiast until whole communities find themselves bettered in ways that no one would have thought possible. Incidents from later chapters illustrate the many methods employed by extension workers in getting farmers to adopt new practices or improve old ones.

Extension is more than an organizational chart and a paid personnel. It consists of the living relationships between the people who carry it on and the individuals who benefit by direct and indirect participation. Most extension services make use of people trained in agriculture, homemaking, or club work, who maintain close relations with the specific groups with whom they work. At the local level the staff members are usually responsible for work in a county, a district, or a region —depending upon the political subdivisions used in a given country. Their titles vary around the world. The agricultural expert, usually a man, is called an agricultural representative in Canada, a county organizer in Great Britain, an agricultural adviser in Australia, an agricultural officer in various British colonies, and a county agent in the United States. Among non-English-speaking countries he is more frequently called an agronome than anything else; such is the case in Denmark, the Balkans, the South American countries, to mention but a few widely scattered areas.

Although the use of women extension workers is not as widespread as that of men, many governments and private agencies interested in rural betterment now realize the strategic importance of teaching farm women better practices in homemaking. In the United States, the county home demonstration workers are assigned to this task, while in Great Britain those thus engaged are called instructresses. As yet no feminine counterpart has gained the international usage enjoyed by the word agronome.

These county or local area workers frequently have assistants. In many countries these assistants usually concentrate on work with 4-H clubs and older youth. The boys and girls carry out projects on their parents' farms and in their homes. Taking care of poultry, raising a garden, or raising a pig or calf is a typical farm project. Making clothes, canning fruit and vegetables, or preparing family meals are typical home projects. The young people learn by doing. The results of their projects are far-reaching, influencing farm and home practices on the home farm and in the community.

The county, district, or regional extension workers are usually backed up by a substantial number of subject-matter specialists. These specialists are the link connecting the research bureaus and laboratories with the local extension workers and the farm people. In general, they are specialists in poultry, dairy, clothing, farm management and marketing, and in particular crops. When working in the county or district the specialist helps the agent put on a demonstration or analyze a new problem about which the farm people are asking the agent's assistance.

The effectiveness of the whole Extension Service is determined in a large part by administrative officers who have little day-by-day contact with farm people but who determine policies and aid in the training of personnel through supervisory services. Most extension work in the field of agriculture is promoted by governmental authority, usually through the Department of Agriculture and sometimes in cooperation with the Department of Education. There are countries, as previously indicated, where much extension has been carried on under the sponsorship of nongovernmental groups such as labor unions, privately organized cooperatives, or religious bodies. Ordinarily, however, agricultural extension, where it does exist around the world, is backed and usually directed by the national government, and extension workers are civil servants.

Practices that have been proved sound by research, experimental trial, and local experience are extended through demonstrations, group meetings, bulletins, radio, press and other media. The strength of extension wherever it is working lies in the fact that it has gained the confidence of farm people by bringing to them tried and proved methods and practices, the value of which they are encouraged to test for themselves. In general, from the findings of science and the experiences of the farmer, the agent helps to develop a plan fully adapted to the local community.

Extension not only takes the findings of science to the farm and there assists in further testing and refining it, but it also takes the problems of the farmer to the research laboratories and helps formulate new types of research activities designed to meet the ever-increasing problems facing farm people. Extension may be thought of as a two-way channel—extending on the one hand the findings of science to the farm people and on the other hand presenting the problems of the farmers to the research specialists for study and analysis.

While living in the midst of a new era of scientific development, most of the farmers of the world still farm and live by tradition. They follow the practices of their fathers and grandfathers, their wives keep house according to precepts that are centuries old, and their families live under health conditions in which survival depends upon folk remedies and sturdy constitutions. In those areas most noticeably influenced by the scientific progress of Western Europe and the Americas, the farmers have begun to replace many traditional practices with scientific procedures, thus becoming a part of a dynamic social order where rapid change is characteristic and to some extent directed. War has intensified this process in many countries. The Chinese peasant in his rice paddy, the Pacific Islander on a pineapple plantation, and the Ecuadorian Indian in his strip of hardy grain fall more and more under the influence of world markets. Where once the weather was the only imponderable a farmer faced, there is now the added imponderable of a complicated price structure which he does not understand and can do little to control. As money becomes commonplace so do other traits of Western society. The technicways begin to replace the folk-ways.

As stated earlier, extension education is based upon the principle of cultural variations and culture change. Every society is unique around the world since it is a product of the geographical and cultural milieu in which it has developed. But there are broad types of societies which have much in common. The nonliterates, for example, by the very fact that they have no written language have to depend upon oral tradition. Accompanying their lack of such a language there is a tendency to develop traits that seem of little importance to one reared outside their way of life, as, for example, the Australian aborigine and his interest in intricate family pattern, or the religious rituals and magical ceremonies of many other tribes.

Peasant people likewise show a number of things in common, especially as regards their general outlook on life. The land is the common denominator which explains their similarities. To be sure, they differ, too, as a result of geographical factors and historical accidents but a later chapter will describe how much alike they really are.

The Euro-American farmer has come a long way from the peasant economy. He is, as a type, under the sway of scientific agriculture and takes a more rational and impersonal attitude toward his occupation. Because of these fundamental differences among the nonliterate, the peasant, and the Euro-American farmer, extension organization and methods will vary; nevertheless, certain basic principles or guideposts, summarized in the last chapter, can receive almost universal application because of the common humanity among farmers everywhere.

While it is important to recognize that the extension approach will have to be adjusted to fit the major cultural patterns of the world, it is also important to bear in mind that within a given country there will be need of variations in methods of organization and the carrying out of extension education. For example, while extension work in Australia is administratively somewhat similar in all six states, it varies greatly among the major cultural regions in methods of organization and working with the farm people. The people living in Queensland, for instance, produce crops adapted to the region, such as cotton and sugar cane, and have many farming and living patterns different from those in the wheat area of Victoria. Extension education in these areas must be and is oriented to the farm and home practices and needs of the area. Likewise in other regions, different adaptations of extension methods are made in keeping with the culture of the area.

As extension work gets a start in most countries it devotes most of its attention to helping the individual farmer improve his farm and home practices. With growing maturity, the extension staff realizes that there is much of importance beside the individual as such. He is a member of a family group and whether or not he cooperates in an extension program often depends upon the reaction of others in his family to the proposal. He is also a member of a local community and the institutions which make communal life possible. Even more, as has already been suggested, he is increasingly participating in the world outside. Since extension accepts education for living as its basic objective, there is developing an ever-increasing challenge to help farmers so to organize that they can participate effectively in their local community and the world outside. If farmers of all countries are to produce abundantly and consume effectively there will be need for greater understanding on the part of all as to the problems faced by rural people in different areas of the world and the essential interdependence of agriculture, industry, and labor.

That the extension job is fundamentally one of education, must not be lost sight of. Education, however, requires more than advisory service to farmers. It must be more than merely presenting scientific facts. Education has a responsibility to society as a whole. It must be increasingly concerned with changes in attitudes, so that people as individuals may progress to higher standards of accomplishment and living in an increasingly complex society.

The contributions of extension teaching in bringing about changes in the way people do things should be evaluated in terms of the contributions made to the social welfare of those adopting the changes. It must also be evaluated in terms of the system of values and beliefs around which the everyday life of any group is organized. By way of illustration, producing more eggs from a flock of fifty hens need not be useful in itself. It generally is useful, but simply because it contributes to the satisfaction and improvement in level of living which human beings, not hens, get out of life. In other words, any innovation in and of itself is not necessarily beneficial; its value depends upon the effect it has upon people and the type of culture change it brings about.

One of the really great contributions of extension education is that it develops people as individuals, leaders, and cooperative members of the local community and the world society. Through participation in extension activities farmers gain a new vision. They are brought face to face with their neighbors' problems and thus aided in seeing the interdependence of their welfare and the welfare of their neighbors, their community, and indeed, the entire nation. Problems are thus recognized as being group problems requiring group consideration and action. Working within the democratic framework which exists in most communities around the world, extension can help farm people not only in the solution of their individual problems but also aid them in the solution of their common problems. Extension then becomes education for action, action on the individual farm as well as group and community action. Experience clearly indicates that farm people when working through a local problem reach a group decision as to what should be done with the problem and act upon this decision.

Such problems as health, education, nutrition, marketing, recreation, and developing interesting and worth-while programs for the older youth are types of activities which require group consideration and action.

Out of extension education the shape of things to come is emerging. All nations will, through the more systematic application of science, realize ever-increasing agricultural production from more fertile soil, ensure happier families in more comfortable homes, greater participation in community life and greater interest and understanding in local, national, and international problems.

Science, wherever it has gone around the world, is responsible for progress in agriculture and rural welfare. Bringing that science to more and more rural people is the job of extension.

Chapter 2 · DIVERSITY AND CHANGE IN THE CULTURE OF NONLITERATE PEOPLES · *By* *Solon T. Kimball*

THE ONCE WIDELY HELD ASSUMPTION that the nonliterate [1] peoples of the earth lived in a kind of cultural "Never-never Land" has receded as we have learned that the basic physical, social, and psychic needs of all men are similar and universal. The study of various societies has shown, however, that there are wide differences in the way in which peoples satisfy their needs. These differences are observable and reveal themselves in the degree of complexity of the technologies, institutions, symbolic systems, and the accompanying customs and beliefs.

The techniques developed and knowledge gained in the study of the simpler peoples have stood us in good stead in the understanding of our own Euro-American society. A wealth of comparative cultural data has been accumulated that gives perspective to our own complexities. In such comparisons we sometimes forget however, that native groups with primitive techniques and unsophisticated ideas frequently have a (to us) complicated social organization and elaborated beliefs and ritual. Many serious conflicts have arisen because of the failure to appreciate the nature of these cultural differences. Only through an understanding of the specific characteristics of particular social groups and of the principles which govern satisfactory relations between them and us, can we hope to translate constructively our knowledge for their use.

Within the time span of recent history primitive peoples occupied the major portion of the world's surface. They were occupants of all the Western Hemisphere, Australia and the islands of the Pacific, considerable portions of north and central Asia as well as the isolated portions of that vast continent, all but a fraction of Africa, and were still found in isolated portions of Europe and the Middle East. The phenomenal spread of Euro-American culture during the past four centuries has swept many of these peoples into oblivion, absorbed others, and greatly modified most of the remainder. The speed with which

[1] The simpler peoples have been variously characterized as primitive, pre-literate, non-literate, pagan, barbarian, native, tribal, and so on. There is no term which is completely satisfactory to designate the peoples whose cultures have developed outside the main streams of Eastern and Western civilization.

this engulfing process continues has if anything been accelerated in recent years.

These contacts have resulted in outright dispossession of the lands of the aborigines and the absorption or extermination of the aborigines themselves; proletarianization coincident with the introduction of a wage economy; and recently, in some regions, protection of economic resources and cultural values with modified acculturation. The character and intensity of contacts have varied greatly. There are, for example, villages in Central America in which half the inhabitants although Indian in origin have become Hispanized, while the others continue the beliefs and practices of an older tribal organization. A few isolated groups such as those of the highlands of the Amazon basin have been relatively free of almost all external influences. There are tribes who, like the Hopi Indians in the United States, have maintained cultural integrity in the face of strong and continuous external assaults.

Sizable areas of the earth's surface continue to be the exclusive homeland of many tribes that are under the real or nominal suzerainty of a colonial power. Nonliterate peoples still occupy most of Africa; great areas in Asia; the desert and grasslands of northern Australia; the Pacific Islands; and the subarctic, boreal forest, desert, and tropical rain-forests of the Western Hemisphere. The remainder of the earth's surface has been appropriated by "civilized" man.

Although environment imposes limitations on man in his use of the materials of the earth's surface, it is also permissive. Through technology, man has been able to adapt himself to all types of physical conditions except the polar ice cap and the extreme desert. Geographical environment is only one among many factors influencing the kind of institutions, customs, belief, and ritual which constitute the major body of the cultural tradition.

Malinowski [2] defines culture as a body of artifacts and a system of customs. The artifacts are made from wood, stone, metal; plants and animals are turned into tools, weapons, utensils, clothing. The system of customs includes the technical knowledge and skill necessary for the creation and manipulation of the artifacts and provides the rules by which individuals and groups relate themselves to one another, and a set of beliefs and practices which relate man to the supernatural. The different peoples of the world, living under different environments

[2] Bronislaw Malinowski, "Culture," in *Encyclopedia of Social Sciences*, IV, 621-645.

and having unique historical experiences, vary widely from one to another in the character and organization of their societies. The fact that side by side in the same environment one may find two different groups, one with a simple technology and social organization and the other an advanced and complex culture, forces us to look beyond environment as an explanation of cultural diversity or complexity.

Complexity of cultures has been shown to be related to the varying number of techniques which taken together constitute the technology.[3] The three basic factors are the type of cutting tools, the methods of food acquisition, and transportation. As there is an increase in the number of techniques, there is a corresponding need for specialization and hence division of labor. As the division of labor increases, there develop procedures which regulate exchange of products, not only within the group but also with other groups. This leads to further specialization, the development of institutions, and the means of enforcing and facilitating the relations among men. From such relationships complex political and economic hierarchies arise. Associated with these structures are value and symbolic systems.

CULTURE CONTACT AND CULTURE CHANGE. Culture comprises an integrated and interdependent whole, which is basically conservative and resistant to change. But people are also adaptable, and, under the influence of new ideas or techniques, changes occur in the cultural pattern which sometimes have far-reaching effects. Contacts established between any culturally diverse peoples are bound to effect modifications in the organization, practices, and beliefs of each. The changes may be gradual and take place over centuries, or they may frequently be rapid and produce extensive and oftentimes disruptive effects.

Culture change comes from the introduction of new techniques, customs, ideas, or practices. The economic, political, or religious pressures which have accompanied or followed the extension of coercive domination by modern imperial powers provide many examples of the effects of culture change stemming from the contacts of those possessing complex cultures with peoples having a relatively simple cultural pattern.

The impact of European civilization on native cultures is dramatically illustrated in many African tribes. The Bantu-speaking peoples of South Africa have been subject to economic, religious, and political

[3] E. D. Chapple and C. S. Coon, *Principles of Anthropology* (New York, 1942).

influences extending over many decades. These contacts have resulted in far-reaching changes in native life, some of which have been disruptive in their effects. Even those tribes which have been least affected have undergone radical changes in economy and social organization.

The Pondo [4] is one such group which has been relatively favored by land resources nearly adequate for the needs of a population possessing primitive agricultural techniques. Nevertheless, the disintegration which comes from the introduction of a wage economy, the migration of natives to mines and plantations, the creation of new material wants, the acceptance of new religious and ethical beliefs, and the imposition of political control has affected the native peoples.

Originally, the Pondo were a tribally organized group, with a system of local area chiefs. Prestige and authority rested in the hands of the elders, in a highly developed ancestor cult. The economy was based on cattle and agriculture. Clusters of households lived together in extended family groupings based on patrilineal descent and patrilocal residence. The members of each group cooperated in the daily and seasonal tasks and shared in the consumption of animal and vegetable products raised by the group. A system of payment and exchange for goods and services had been developed which covered all aspects of life and included a complicated pattern of exchanges and obligations in securing a wife.

The chiefs exercised nominal authority and had important responsibilities for interpreting customs, settling disputes, regulating relations with other tribes, leading war groups, and performing the ceremonies in connection with bringing rain to break droughts. Status was less dependent upon wealth than upon generosity, and the members of the group maintained close relations through continuous exchange of gifts and services.

European contacts have effected changes in all aspects of Pondo life. The major influences have come from the imposed political domination, employment away from the reserve on farms and in mines, and from association with missionaries, educators, and traders. The trader supplies goods which are paid for with money earned in employment. The increased dependence on trade goods has resulted in a decline of native handicrafts. The importance of money and a wage economy extends into the system of exchange and payment, fostering the development of class distinctions based on wealth and education.

[4] Monica Hunter, *Reaction to Conquest* (New York and London, 1936).

Wealth rather than generosity has become the basis for status. There is an increasing economic individualism and consequent decline of the interdependence of the extended family grouping in production and consumption activities.

The decline in the prestige of the elders and adherence to the ancestor cult is correlated with the economic changes. The young and vigorous are the ones who seek wage employment and return to their homes with the money, which gives them status and independence of their elders. The schools and the missionaries teach new ways of doing and believing and cast doubts on the efficacy of the old practices. The Christianized natives no longer look to the chiefs to perform the ceremonies to break the droughts, but hold prayer meetings in their churches. Parental control over the young has been weakened and even the rigid practices governing the family and sexual rights have changed. Women are no longer dependent on the status of wife to provide their material needs.

New social groups are emerging, based on religion, occupation, and training. A growing nationalism is accompanied by a racial and political bias, with criticism and dislike of Europeans and things European. This has found expression through the African National Congress, native Bantu Churches, and a Bantu Trades Union. These latter organizations have also found support among the natives who have left their homes for residence on the farms and in the towns. These emigrants face even greater difficulties than do those who have remained on the reserves. Since the natives are barred from owning land, their only means of support is to work as farm servants. If they are dismissed and have lost their claims to the reserves, they become completely destitute. Although some of the tribal sanctions continue to operate among the class of farm servants, these are divorced from the tribal setting and tend to become meaningless. The effect on the town dweller is even more marked, and tribal sanctions disappear in the proletarianizing environment of dependence on wages for a livelihood.

Significant changes in the cultural pattern are not necessarily dependent on the aggressive contacts between a complex culture and a simpler one. A people may adopt a practice freely and without pressure. Such adoptions, however, are most likely to occur in technology. The Tanala of Madagascar [5] provide an interesting example of the profound effects which a modification of techniques has had on

[5] Ralph Linton, *The Study of Man* (New York, 1936), pp. 348-354.

the cultural pattern of a people. Over a period of several decades the Tanala changed from dry-rice to wet-rice cultivation. During the period when dry rice predominated, they lived in villages which were moved every five to ten years as the cultivated land was exhausted. The villages were independent units organized on a democratic joint-family basis with assignment of land based on need, cooperative labor, and equal distribution of the produce. There were no classes, no private ownership of land, and no marketable surplus.

The introduction of wet-rice cultivation made possible the continuous farming of one tract, but since there was insufficient land that could be irrigated, it was necessary for those with dry-rice cultivation to move when their land was exhausted. Wet rice brought with it the development of rights in landholding and no redistribution on the basis of needs, a class of landholders, the increasing importance of the household and lessening importance of the joint-family, both in ownership, cooperative labor, and sharing. Village life became permanent. Villages were strongly fortified, and groups of villages became associated as tribes and later in a kingdom. Slavery was introduced and class differences based on wealth and land ownership became marked.

Contrast between new and old in social organization and techniques is provided by one of the Tanala clans which originally accepted wet rice, but later rejected it and has persisted in preventing the introduction of wet-rice cultivation into its area by its own people or by others. There the pattern which originally obtained throughout the area of Tanala occupancy continues.

It is improbable that an administrator could have gone among the Tanala and, after observing the social organization and customs surrounding the use of land, production, consumption, and so on, could have consciously proceeded to produce the results described above. It is almost certain that if such a hypothetical person had appeared and contended that land should be privately owned, that the individual household should be the primary economic unit, that there should be status based on wealth with social classes, and that the adoption of wet-rice cultivation would bring about all these changes, he would have received little understanding and no support. Nevertheless, if he had demonstrated the advantages of wet-rice cultivation, it is probable that this technique would have been adopted, with all the results cited.

The American Southwest provides us with another example of the effects of culture contact and resultant culture change. It is significant because this area was inhabited by a number of culturally diverse peoples, subject to similar external cultural influences and yet exhibiting wide variations in the type and extent of culture change arising from culture contact.

While not all were exposed to the same degree of cultural intrusion, there were some tribes which accepted and modified certain elements readily, but others exhibited entirely different tendencies. The two streams of Euro-American influence provide further complications in evaluation. From 1540 to 1845 the Spanish-Mexican influence was predominant. Since that time Anglo-American influence has been gaining ascendancy, although for the Pueblo dwellers the Spanish elements continue strong.

Culturally, the least complex of these tribes were the Piutes who lived in the Great Basin area beyond the main currents of Spanish contacts. They lived in small bands moving from one locality to another gathering seeds, plants, nuts, and insects, and killing small game. They possessed chipped stone tools, built temporary brush and skin shelters, and wore few clothes. Much energy was devoted to the wresting of a precarious existence from an unfriendly environment. The simple ceremonial life was observed chiefly within the family or during the winter months when several bands came together. The Piutes were soon crowded out of their area by the incoming American settlers.

At the other end of the cultural scale were the Pueblo dwellers of the Rio Grande Valley and the Zuñi and Hopi to the west.[6] These people were irrigation agriculturists, made pottery, wove cloth, and built houses of stone. They had a complex social organization, with theocratic control exercised by a priestly group. Although the Spanish converted them to nominal Catholicism, they remained basically native in their culture. Certain agricultural techniques, some tools and plants were accepted, but, unlike the Navajo, the Pueblos continued to rely upon agriculture for food. Domestic sheep and cattle, although acquired in small numbers, never became of basic importance in the economy.

There was mutual adaptation and assimilation between many Pueblo towns and the Spanish with intermarriage and cultural blending.

[6] Pueblo peoples originally occupied a much wider expanse of country; archaeological remains at Chaco Canyon, Mesa Verde, and other places attest to their architectural and other abilities.

Pueblo organization was not only enriched but it also gave much to the newcomers. It is more difficult to assess Anglo-American influence because the process is still continuing. It has not been great except through the schools, because the Pueblos found little of value in the new culture.

In southern Arizona live a number of other groups, the best known of which are the Pima and the Papago. They practiced irrigation agriculture and possessed a technology somewhat comparable to the Pueblo. The nature and effect of Spanish influence did not differ greatly from that with the Pueblos. The Anglos have restricted the area of occupancy, introduced schools, and attempted to instruct in methods of agriculture and animal husbandry, but without too much success.

The east-central portion of the Southwest was inhabited by widely scattered Athabascan speaking peoples.[7] The early relations between these groups and the Spanish differed from the Spanish-Pueblo relationship. The Apache and Navajos were a constant source of trouble and friction, and frequent expeditionary forces were sent against them. They quickly adopted the horse and the tools and implements of the hunt and war. The cultural influence was probably greater on the Navajo, who learned metalworking, adopted some plants, and were tremendously influenced by domestic livestock which they quickly adopted. Extensive culture change resulted. The examination of these changes and the recent attempt of the Indian Service to regulate the use of land and livestock constitutes an informative chapter in culture change and resistance to external pressures.

The Navajo today number 50,000 occupying an area of nearly 25,000 square miles in Arizona, New Mexico, and Utah. Their homeland is rugged country of semidesert and steppe terrane. It is broken by deep canyons, high plateaus, and a limited mountain area covered with coniferous forest. At the time of Spanish contacts in the fifteenth century, the Navajos had a simple agriculture, but also depended heavily for food on game and nuts, seeds, and grasses. The acquisition of the sheep and horse from the Spanish opened the avenue for a radical shift in technology and brought about changes in social organ-

---

[7] These included the Jicarilla Apache of northern New Mexico, the Western Apache of east central Arizona and western New Mexico, the Mescalero Apache of central New Mexico, and the Navajo Indians of northwestern New Mexico and northeastern Arizona.

ization, mythology, ritual, population spread and density. Within a few years dependence upon these animals for food, clothing, and blankets was to replace the previous dependence on wild game.

The horse permitted the Navajo to extend his quests for wild game and plants over a much greater territory and also allowed the quick transport of food to permanent or temporary camps from much greater distances. The need to take the sheep to water and pasture by day and to protect them from human and animal marauders at night demanded a permanence of residence but also necessitated seasonal migratory movements to new feeding grounds and water supplies.

Competition from livestock eventually destroyed the habitat of the wild animals, and with their disappearance the full dependence upon domesticated animals was completed. The sheep and horse also harvested the wild seeds and grasses which had formerly supplemented agricultural production, so that food gathering declined to little importance. The prayers, ritual, tools, and techniques associated with hunting have nearly disappeared. There is, however, a growing body of knowledge, myth, and prayer connected with sheep and horses.

The horse, by extending the range of movement, and the sheep, by creating the need for seasonal movements and for migration to areas relatively ungrazed, both permitted and enforced the permanent settlement and utilization of formerly unoccupied areas. The direction and tempo of this expansion was influenced by an increase in population, an increase in livestock numbers, and external pressures. By 1900 the area now known as the Navajo reservation had been completely occupied.

The security against enemies provided by external control and the introduction of animal husbandry produced some changes in social organization. The Navajos are a matrilineal, matrilocal people. Property theoretically passes through the female line. The extended family is an effective cooperative producing and sharing group, with several of these units comprising a band. The band is composed of persons related by blood or marriage occupying a recognized territorial area, cooperating and sharing in certain common problems, and with a fairly well-developed leadership.

Significant changes appeared in the widened spread between the rich and poor, livestock became a basis for status, inheritance through the male line became more common, and claim to given areas of country on the basis of occupancy and use was strengthened. The ownership of many livestock was desirable. A rich man would have influence and friends. He need never be hungry or ashamed, for the sheep would provide him with food, and he could sell the wool and animals to obtain clothing and jewelry. With these values and incentives, it is easily understood why great efforts were made to increase the numbers owned.

Navajo society is an expanding one. The area of occupancy has been greatly increased in the last two centuries, as has the population and the wants of the peoples. Under a static technology, it was necessary that the expansion be equal in all directions. The inability to extend their grazing led not to a reduction of wants or diminution of population increase but to an overuse of the range by excessive numbers of stock.

The shock of drought and depression in 1932 disclosed the fundamental weakness of Navajo economy and technology. Drought revealed the magnitude of overexploitation of the range resource with consequent erosion and loss of plant cover. Depression revealed that the increasing wants of an expanding population could not be met by the current technology unless commensurate territorial expansion was possible.

The efforts to correct the situation through reduction of livestock numbers has stimulated strong opposition among Navajo and trader. When it is remembered that livestock formed the central core of the technology through which the needs for life were secured and status was achieved and expressed, it is easily understood why reduction measures were so strenuously opposed, for any interference with the ownership and use of livestock threatened the entire economic, social, and religious systems of Navajo society.

EXTENSION PROGRAMS AND NATIVE PEOPLES. Many native peoples are facing situations similar to that of the Navajo. Many governments are concerned over adjustments that must be made if their wards are to survive. These changes can be made much more readily if those who are the administrators, the planners, and the executors understand the nature of culture and the process of culture change.

The era of unplanned and unregulated exploitation by native peoples of their area of occupation has been greatly curtailed in recent decades. Free economic exploitation by outsiders has also been limited. The controlling governments have been gradually evolving policies designed to ensure protection of the aborigines. The United States Indian Service during the past decade has pursued an aggressive program which has halted the alienation of Indian lands, added additional acreage through purchase and consolidation, and instituted policies for conservation of forest, range, and farm land. Through its extension service, improved techniques for handling livestock and farm land have been introduced.

Several of the Central and South American countries have recently embarked on similar programs in connection with their Indian groups. Colonial officials of other great countries have also shown increasing concern. Some of the administrative problems of the British in Africa have been worked out through an "indirect rule" policy which permits a measurable degree of local autonomy in settling tribal problems and protects the natives from the less desirable effects of imposing on native groups European notions of family responsibility, property, inheritance, and the like. Programs of education, health, and conservation of resources are also in operation in several areas, and administration officials are endeavoring to protect the natives from uncontrolled exploitation of their reserves by nonnatives.

The production of food and other vegetable products has become of great importance during the present war, and we may expect that native peoples will continue to supply certain types of products for the world markets and foodstuffs for garrison troops after the war. Local administrators carry a heavy responsibility lest these native groups, thrown into the system of world economics and exchange, do not become demoralized and eventually an indigent proletariat.

Extension workers with a knowledge of the principles of cultural anthropology can make a fundamental contribution among native peoples. Field workers will have the greatest success if they utilize the existing social organization in introducing or modifying techniques, some of which may require acquisition of knowledge and skills foreign to the group. They can protect the natives from the disruptive influences which too frequently follow abrupt changes in the relations within the group or between the group and Western peoples. One should remember that the techniques, institutions, beliefs, and practices represent an integrated and functional whole and that changes in any one element will have its effects elsewhere. Effective changes in culture occur slowly, and any efforts threatening the established institutions which provide status and regulate human relations will be resisted.

Chapter 3 · EXTENSION WORK IN THE PACIFIC ISLANDS · By Felix M. Keesing

INTRODUCTION. The Pacific Islands offer a fruitful and in many respects urgent field for extension activities. The peoples "native" to these hitherto isolated islands are being swept, whether they like it or not, from their circumscribed traditional settings of life into a larger world. In some areas the changes have been so telescoped that representatives of the living generations run the gamut from the former head-hunting warrior to the college-trained youth. Their lands have been invaded here and there by white traders, agricultural settlers, and miners, and by Asiatic laborers, merchants, and moneylenders. They feel the impact of booms and depressions, and now of a war that has turned many of their homes into battle grounds.

World War II is having tremendously disorganizing effects, both among the peoples caught in Japanese occupied territories and among those in islands mobilized as United Nations bases. Of the Solomon Islands, for example, an official writes early in 1944:

These islands have reached rather a critical stage in their development—war has caused much bewilderment in the minds of the people, and a certain amount of economic distress. The presence of soldiers has accustomed them to a higher standard in the possession of luxuries, and has placed them in a position of extreme inflation as regards their own native exchange (e.g., bride-prices). The retreat of commercial enterprise has led to the inauguration of a considerable amount of government commerce, which will undoubtedly increase and may even replace private enterprise to a considerable degree. . . . In other words the whole economy and most of the society is being changed and reorganized.

Postwar readjustments, and considerations of a long-term policy in such areas, are going to call for the most expert help available.

Magnitude of the Problems. A preliminary idea of the magnitude of extension problems in this region may be gained from a brief summary description. The tropical islands of the open Pacific, often spoken of as the South Seas, have a combined land area of about 400,000 square miles. They range from huge New Guinea, more than twice the size of Japan proper, to hundreds of tiny specks. Together, they have a population of over two million brown and dark-skinned islanders: the so-called Polynesians, Melanesians, and Micronesians.

If the Malaysian islands farther west are added, the total area rises to about 1,200,000 square miles (over one third the size of the United States), and the native population to well over a hundred million.

Extension activities must take constantly into account the great diversity of conditions in this island region. The scattered archipelagoes and islands, through the play of imperial forces, have become subdivided politically into no less than twenty-five separate jurisdictions—colonies, territories, protectorates, mandates, and others—controlled by nine different Powers. In turn, within each jurisdiction, the island peoples are often racially diverse, and are broken up into local groups that differ in language, economic base, religion, and other matters of custom. In Malaysia, many native states have consolidated, but more than three hundred exist today, large and small, ruled within the colonial framework by their sultans, rajahs, or princes. But, for the most part, the people have been broken up into little politically autonomous groups, sometimes each settlement traditionally standing on its own. The modern colonial units are obviously artificial constructs, and political consolidation, as in the Europe of earlier days, moves slowly.

Modern immigrant groups—whites, Chinese, Indians, and others—tend to concentrate around the small number of urban centers, nearly always ports, and in the relatively few well-developed plantation and mining areas. Roads and other modern facilities, too, are limited almost entirely to these special localities. Those islanders whose traditional home happens to be here and those who have come as permanent residents are most affected by the money economy and other Western influences. Native labor is also drawn temporarily to such districts. But the great majority of islanders continue to live in their scattered ancestral communities more or less on a subsistence basis. In the upland fastnesses of New Guinea and some of the large neighboring islands, there are still isolated groups who are outside government control and may never yet have seen those vanguards of white civilization, the trader, the official, and the missionary. Extension planning must take into account all these special human dimensions: urban natives, laborers, natives of the outer districts, and the nonnative groups.

Extension Work to Date. Work more or less equivalent to extension activities has been carried on by colonial governments for many years, through land, labor, agricultural, and other administrative departments, and also to some extent by missions and private industrial organizations such as Planters' Associations. Colonial conditions have been a forcing bed for many such activities, especially as regards government participation.

Decades back, governing authorities usually of necessity had to introduce measures regulating land transactions between natives and nonnatives, recruiting and employment of labor, plant and animal quarantine, and many related matters. Both governments and missions have interested themselves increasingly in native economic welfare, while private industrial and commercial interests have joined with governments in programs of research and experiment, pest control, and similar activities.

The results of such work to date have been very uneven, particularly as regards native welfare. At the earlier stages, in fact, most official programs were usually focused upon getting suitable areas opened up for nonnative exploitation—settling land titles, securing a labor supply locally or from overseas, and so on. Most of the measures touching the natives were protective, designed to prevent their being exploited or crowded off the land, and so becoming an embarrassment on account of poverty and even active hostility. Later, the governments assumed more ameliorative programs, based on considerations of positive welfare and the increasingly serious problems that emerged as the islanders widened their contacts with modern civilization. Among such problems have been population increase—even in some places serious overpopulation; a growing incidence of tenancy, usury, and poverty; unemployment; and inadequate nutrition.

Two main phases of native economics have received emphasis. First, the authorities have wanted to reinforce and expand the traditional systems of food production, so as to assure the greatest possible stability in local subsistence. Second, they have tried to stimulate the development of commercial enterprises by native producers, so as to educate the islanders in the modern economy and add to the prosperity of the territory. Some governments have gone so far as to compel native families and communities by law to plant certain minimum amounts of crops annually. Others have offered prizes or inducements, as in Papua, where natives are excused from paying taxes if they develop plantations under official supervision, and return part of the revenue to the government in lieu of taxes. Most governments have model plantations and experiment stations designed to stimulate native agriculture, and a staff of advisers and inspectors to make the rounds of the native communities. Increasingly, competent islanders are being trained for such tasks, in some jurisdictions up to the college level, and this is proving one of the keys to successful work.

Official encouragement of native commercial production was particularly vigorous up to 1930. Then depression swept like a hurricane through the island economies. With markets glutted, prices plummeting, trade stores closing, and many nonnative planters driven to the wall, bewildered native groups were forced back largely upon their older modes of living. Government policies shifted toward reinforcing local self-sufficiency, and developmental work had to be curtailed because of shrunken budgets. Throughout the thirties, demand for virtually all native-produced goods continued to be limited and uncertain, while nonnative industry had mostly to be bolstered through international quota systems, imperial subsidies, and other artificial means. Since 1939, war has given a fillip to production in many areas, though in Japanese-occupied zones numerous groups have been thrown back entirely on their own local resources as arteries of trade dried up. The significance of all this for rehabilitation programs, and especially for long-term extension policies, will be discussed in due course.

NATIVE ECONOMIC SYSTEMS. The principle that extension programs must be built upon the realistic conditions existing in any locality and upon the modes of life of the people being dealt with is demonstrated with particular clarity among the island groups, which differ so greatly from one another.

Perhaps the first outstanding characteristic of economic life to be taken into account by the extension worker in such an oceanic setting is the maritime or aquatic nature of the environment. The vast majority of the islanders have their settlements where they can get access to the sea and its resources, and even the inland peoples live mostly along the lakes, swamps, and waterways. Tropical shores, reefs, lagoons, and streams are rich in plant and animal life, and many peoples are, so to speak, intensive aquatic farmers. Here a whole new realm will be opened up to the extension worker, especially as regards the use of vegetable products of the sea. He will in fact be pupil rather than teacher, except as his ingenuity and his microscope may enable him to improve on the age-old skills of a locality, or he may be able to pass on to one group knowledge from some other group.

Some of these native economies represent a highly specialized adjustment to a limited environment, and here extension programs would have to be based upon very precise local knowledge and study. On the more barren coral atolls, for example, such as the Gilberts, the first reaction of the outsider may well be of wonder that any human group could settle there permanently at all. Then, as well-populated villages are seen—the Gilberts actually have almost 200 persons to the square mile of land area—and the good physique and splendid teeth of the people are observed, it becomes apparent that the limited range of land products such as coconuts, pandanus fruit, and a coarse species of taro which can be grown in saline pit water, together with fullest exploitation of sea products, can yield a satisfactory livelihood.

In a similar way, a specialized economy may have been built up by swamp peoples, particularly through use of flour from wild or cultivated sago palms as a staple, and by peoples of barren mountain heights through intensive terraced gardening. In the great rain forests of Malaysia and New Guinea there are still scattered nomadic bands of hunters and gatherers of wild plant foods, shy peoples who are sometimes of pygmy size. Except as such groups may prove willing to migrate to places with more diversified opportunities—a proposal which so far has nearly always been vigorously resisted—the extension worker has here very narrow if interesting fields for planning. In general, these isolated groups are the most conservative among the islanders, as they see relatively little in the white man's civilization that can be of use to them. One obvious, if rather negative, contribution that extension activities can make is to help exclude from their areas any pests or diseases which may threaten their limited resources.

By no means all island settings are so restricted. Though tropical soils are often very poor because of leaching and other factors, some valley and lowland areas have good alluvial soils, and here and there a particularly rich area has been produced by volcanic activity, as in fertile Java. Even the less favored sections may produce luxurious crops for a season or two, and perhaps for an indefinite period if drained, irrigated, and fertilized. Peoples living on the higher islands, especially on firmer ground back of the frequently swampy coastlines, may therefore have wide scope for agricultural development.

At the one extreme, under such conditions, are peoples like the Javanese, or the Filipinos of the Ilocos coastal strip in the northern Philippines, who have intensive systems of cultivation. Their economic activities are concentrated primarily upon the annual drama of winning from their limited fields sufficient amounts of high-yield crops of paddy rice, corn, and cassava, to feed the teeming and ever-increasing population. Water buffaloes or oxen are used along with hand labor to prepare the soil; and traditional methods of cultivation, improved under modern conditions particularly through government aids, maintain a level of high production. Farther east, some of the South Sea peoples are very meticulous in growing their staple root crops: yams, taro, and sweet potato.

More typical of native economies, however, are much less intensive kinds of land utilization, in which a simple digging stick is the standard implement. Rainy hill slopes may be used to grow "dry" rice or taro and other easily cultivated crops, which, along with tropical fruits like coconuts, bananas, and breadfruit, grown in the gardens or around the villages, and wild products of the forest and sea, provide an adequate economic base. A characteristic type of cultivation is a so-called predatory or shifting agriculture, or "fire farming," by which the group concerned cuts and burns a new clearing every season or so for its gardens. In Malaysia and the New Guinea region many inland groups practicing this type of cropping are seminomadic; their settlements consist of small hamlets of a few families each, dotted sparsely through the forests.

Efforts to date have not been particularly successful in getting such predatory cultivators to settle down or to practice more intensive agriculture, though sometimes missions and governments have managed to consolidate scattered hamlets into larger villages. From the viewpoint of the native, crops can be grown with minimum effort in this migratory pattern, and land has usually been plentiful. To the authorities, however, the cropping method is anathema, because of progressive destruction of valuable primary forests and also because of the persistence of native ownership titles over such apparently abandoned holdings and the trouble that is likely to ensue if settlers from outside attempt to occupy them. As a compromise measure, some agricultural departments are trying to get the native groups to plant useful trees—such as rubber or teak, supplied as seedlings from government nurseries—in the burned-over clearings instead of letting them become choked up with useless second-growth jungle or spear grass. In such areas this is obviously a major problem challenging the extension worker.

In addition to food crops, the islanders grow, or collect in the wild state, numerous other useful plants—grasses, leaves, and fibers for thatching, weaving, and other purposes, medicinal plants well worth careful study by local health staffs, fish poisons such as derris (rotenone), and perhaps narcotics (betel and kava root). Native communities nearly always have domestic livestock, which may include pigs and chickens. These may provide the main sanitary system and, among more conservative groups, may be used for animal sacrifices, omen reading, and ritual feasts. Some peoples raise fish and other aquatic products in salt- or fresh-water ponds, and in water-flooded fields along with their crops. Some groups are able to supplement their diet with wild game, including hogs, deer, birds, and even certain reptiles, insects, and other unusual edibles. The forest and sea can also provide a wide range of emergency foods that will sustain life, as Allied troops in the South Pacific have learned. The extension worker would do well to know not only the practical value of all such products, but also what they mean to the people with whom he is dealing, especially any taboos or other religious significance pertaining to them.

Some of the more isolated communities were wholly self-sufficient before the coming of the trader in modern days. Nearly everywhere, however, the islanders engaged in at least minor trade or barter, or its equivalent in ceremonious "gift exchange," so as to round out their needs. Coastal peoples usually exchanged their maritime products for forest products controlled by the interior peoples. Here and there a group controlled some special resource such as clay for pot-making or a salt spring, or else had some unique skill like building extra-seaworthy canoes. A few groups even won their living to a considerable extent as middlemen along the coasts, rivers, or main trading trails. Much of this native interchange continues today, and has important social as well as economic significance.

Yet, paradoxically, natives have rarely taken up modern style storekeeping or other commercial occupations, leaving these in the hands of whites, Chinese, or other outsiders. The extension worker, discarding the hasty judgment of some observers to the effect that the native "lacks the commercial instinct," will be able to weigh the economic and other values in native life that often make it virtually impossible as yet for a native community to accept one of its own members as a merchant, or that bring quick bankruptcy to any individual rash enough to try to buy and sell among his kinsmen and neighbors.

MODERN CHANGES. Away from the few ports and industrial areas, native life is likely to show surprisingly few outward signs of change. House furnishings may include oddments of Western gear, including kerosene lamps and enamel basins; former Stone Age peoples will have metal tools; garments are now nearly always made of cloth, though cut to native styles; certain outside crops and animals have been adopted. Closer scrutiny, however, usually reveals that the traditional native economies have been considerably, even if selectively, modified. The extension worker will want to delve into the background of such changes for the lessons they offer in relation to his own planning.

Some modifications have been made voluntarily by the people as a result of being faced with new objects and opportunities. They represent a process of self-motivated education of the sort that extension work tries to stir into activity and harness to its enterprises. Thus all natives, except those few beyond the range of the trade store or itinerant trader, have entered some distance into the money economy. The most practical index here is an inventory of any local trading establishment to determine what commercial goods a particular group has come to count as necessities—matches, cloth, soap, kerosene, and so on—and also what they like to buy in good times or for special occasions as luxuries. To get such articles, natives must either produce surpluses of goods in commercial demand or sell their labor; in some instances they are able to rent land or collect royalties from nonnatives for use of resources, but this is proportionately infrequent. Other changes are not self-motivated, but have come about through government compulsion, as with the imposition of taxes or labor services, or perhaps as the aftermath of conversion to Christianity, through urging by missionaries. All this brings about far-reaching if often subtle modifications in native economic organization and values.

At least equally important are the negative aspects—what elements of modern experience have been available, yet have aroused no spark of interest or have been actively resisted and rejected. A group which gives generously to church collections may be quite unresponsive to the idea of lodging money in a bank. Or a people may have resisted the introduction of latrines by getting out their spears. An extension worker might quickly see why. Human excreta and other waste here perhaps went into the pigpen, and pigpens supplied essential fertilizer for the fields which fed the people: to have latrines would break this tight cycle. Indeed, the fact must be faced, that many natives today feel that they have picked up by now about all the tricks the white man is capable of teaching them; the result is a tendency to lapse into a new conservatism based on the fusion of native and alien ways which has been worked out in recent decades. To set in motion dynamic changes again in these conservative matters that touch upon food and other essentials of physical survival will call for doubly convincing kinds of demonstration and education.

The selective nature of native adaptation is well-illustrated by the types of product which such groups supply to world markets. Commercial crops are almost wholly limited to those which occupy only a small part of the land and which require a minimum of labor, care, and technical skill. The islanders are, of course, virtually ruled out from growing products that call for large capital investment, elaborate processing on the spot, or very exact standards of quality. Copra (dried flesh of the ripe coconut) has long been the principal item of native commerce in the islands, sometimes supplemented, where markets are available, by tropical fruits like bananas and oranges, or by yams, cassava (tapioca), pepper, kapok, and a few other crops. In Sumatra and other places in western Malaysia an increasing number of natives are growing rubber and coffee. Coastal peoples may gather trepang (*bêche-de-mer*, dried sea slug) for the China trade, commercially valuable shell, and in a few areas still, pearls; while forest peoples may contribute gums and resins, rattan, ivory nuts, and a few other products. But markets for most wild products of this kind have fallen away in recent years, particularly through competition from synthetics. Some groups make and sell hats, mats, and curios.

Native output tends to be sporadic, depending on the immediate needs of the people for trade goods or money. The Westerner is puzzled at first by the perfectly logical fact that production tends to increase in quantity when prices are low and to decrease when they rise, so the native's modest needs can be met at the lowest cost. But there is a slow over-all trend upward as the islanders enter increasingly into the commercial economy. Governments have found to their cost that overstimulation of native commercial production is likely to be disruptive rather than beneficial, as it puts money or luxury goods into hands of natives beyond their effective needs. In consequence, gardens and fishing grounds may be neglected in favor of canned foods, natives may crowd into the ports, or other undesirable or dubious results may be forthcoming.

During the lean years of depression and Japanese occupation, communities tended to be thrown to the opposite extreme. Though having the backlog of local subsistence, they have experienced a definite lowering of their levels of living. In the case of urban natives, and those few who had moved further ahead in commercial economy, the pinch has been extreme. The sudden arrival of thousands of troops in many formerly isolated districts, bringing hitherto undreamed of goods, having money to burn, paying fabulous prices for laundry and other services is bringing economic change and disorganization on a fantastic scale. The extension worker in such areas in the future will certainly have his hands full trying to piece together a stable economy.

PERSISTING NATIVE ORGANIZATION AND VALUES. The diverse economic systems reviewed above, modified more or less in the course of modern contacts, are rooted in persistent modes of organization and thought that can only be understood to the extent that the worker turns ethnographer. Here lie pitfalls to be avoided, but also the vital springs which, rightly touched, will release native energies in the service of extension programs.

Methods of production, distribution, ownership, and consumption are invariably complex, even in what appear to be the simplest island societies, and the economic philosophies have all kinds of subtle nuances. (The extension worker will doubtless often wish that the naïve picture of Karl Bucher's "pre-economic" savage were true.) Even when a group consists of what appears at first to be a few dozen people, native belief may hold that the ancestors and spirits are present in force, so that the effective group for economic and social, as well as religious, purposes is reckoned at least in hundreds.

Obviously these island peoples differ greatly from one another in what they count as wealth, in the wants they will expend effort to supply. The white settler is often critical of natives for their supposed "dumbness" as regards commercial opportunities available to them, or the way they "waste time" on "useless" activities like accumulating ceremonial wealth such as shell money or carrying on elaborate rituals and festivals. The fact is that native conceptions of "the good life," their standards of living, continue to be expressed primarily in terms of local resources and traditional habits. Our measures of economic worth, whether a bank account or up-to-date plumbing, are likely to cause hardly a ripple among them. However strange, and at times crude, they may appear, the native value systems must form the base line for extension activities.

Invariably, a close connection exists between the important economic pursuits and the local religion. Producing the staple foods, building houses, and everything else vital to survival and welfare are saturated with religious ideas and rituals, with magic and taboo. It is in these activities that the native peoples undergo some of their greatest difficulties and anxieties, and therefore seek aid from spiritual forces as they interpret them. Even among groups converted to Christianity or Islam, the familiar local procedures of ritual and magic tend to persist as part of the essential "technology" of economic operations. To get a people who practice shifting cultivation to adopt a sedentary form of life would call for spiritual as well as material rehabilitation, and any innovator has to be willing to take supernatural risks. The rich feeling that native groups have for their soil, sea, and sky, described by some observers as a "mystical communion," need not be wholly a deadweight as regards extension programs, but may sometimes be harnessed by the skillful worker to give drive to his plans.

The economic life of the islanders also ties deeply into persistent social customs which must be understood and utilized. Household, kin, and neighborhood groups, and even whole communities work together in various productive enterprises. Within such groups, too, there is economic specialization on the basis of age, sex, class, and other phases of the social structure. It would be difficult, for instance, to get men to work in spheres that traditionally pertain to women. Goods pass from hand to hand in marriage settlements, as gifts to visitors, and in many other social transactions. The status and prestige of sultans, chiefs, and other leaders, and of the aristocratic classes, is reinforced by the accumulation of valued property, the giving huge feasts, and other ostentatious handling of wealth.

The extension worker must know thoroughly the various types of leadership and authority in any community: heads of households, kin elders, religious leaders, native officials exercising powers bestowed by the imperial governments, and so on. The success of his enterprises will largely depend on his ability to get along with, and use such leaders. In some instances, by attaching prestige value to something he wants to introduce and by getting an influential person to sponsor or adopt it, or even by giving it ceremoniously as a present to such a person, the extension worker may rapidly create a widespread demand among an otherwise conservative people. At the same time, he may have to circumvent a tendency for any such leader to create for himself or his class a monopoly over such an innovation for prestige reasons.

Some of the greatest successes of the extension worker may come if he is able to harness to his program the important cooperative institutions existing in all these little intimate island communities in greater or lesser degree. These peoples, for example, are generally used to working in groups, rather than as individuals, and a native working bee can get prodigies of labor done if the right incentives are there. Native work habits and values are very different from those of Western countries, with their clock-and-calendar regularity, high degree of specialization, and sharp separation of "work" from "leisure." The islanders take part in many different kinds of work, timed to the broken rhythms of seasonal and weather changes and the sudden emergencies which life close to nature presents. The drudgery aspects of work are often relieved by singing, gossiping, feasting, and other accompaniments which rally morale and make the task as far as possible an enjoyable social activity. Major enterprises are usually stamped with the solemnities of religion to rally both human and spiritual forces.

Cooperative usages also tend to dominate in property holding. Some resources and goods will be public property, just as in Western countries, while others may be held by kin groups, households, and other social units, as well as by individuals. Customs here are frequently complex, and it may be best to forget Western concepts of "ownership" and "rent," asking instead, Who has authority over the piece of land? Who uses it? Who has rights, interests, equities, responsibilities in relation to it? Where multiownership occurs, it is not some vague form of communalism or collectivism as some have supposed; the rights of all individuals concerned are meticulously defined by the local customary laws on property. Serious misunderstandings and conflicts have emerged in some places where white settlers have thought that they have bought property when all they had done was to pay for one individual's right in it.

Inevitably, in the modern setting, there tend to be strains and stresses as the older cooperative institutions are impinged upon by newer ideas of rugged individualism in property holding and economic enterprise. A small but increasing minority of islanders are adopting the white man's values. These are often local chiefs or aristocrats—who then tend to metamorphose into landlord gentry—and younger people who have been to advanced educational institutions. How far the extension worker will be justified in pushing individualism, and to where he may find adequate opportunities for persons so marked, without undermining the cooperative institutions valued by others, will have to be judged through close study of local circumstances. Some students of native welfare have hoped that communities could be steered around the worst phases of competitive individualism by linking cooperative values directly to modern enterprise. In this connection a promising growth of native cooperatives in the Gilberts and a number of other island areas deserves scrutiny.

The island peoples are sometimes accused of being improvident and happy-go-lucky, living only for the day. One contributing factor to this impression is the lack of storage facilities to carry over surpluses and so give greater economic stability. It must be remembered, however, that many native products deteriorate very rapidly in tropical, insect- and bacteria-ridden settings. The extension worker will find that the native is likely to adopt a system of live storage, so to speak, leaving perishable products in the garden, in the forest, or in the sea until such time as they can be used. Nevertheless, some peoples have worked out means of preserving foods by drying, smoking, fermenting, and other techniques, and of storing seasonal crops of grains and tubers. In the interests of widening the margin of security, extension programs can justifiably give considerable attention to this problem of storage, encouraging the transfer of current native techniques to groups not acquainted with them and developing new ones through local experiment.

Native handicrafts and industries may also come importantly within the scope of extension activities. Each island people has its own traditional crafts—objects of wood and metal, weaving, making shell ornaments and so on. Upon some of these is lavished the extra attention that turns them into works of art, especially if the objects are associated with religion. In modern days, some native industries have declined or died out, while others have persisted and may even be enhanced through commercial or other demand. New types of craftwork have also been developed; these include hatmaking and the carving of Western-style objects. In a number of areas, the government has helped with industrial training and with the improvement and marketing of craftwork, so as to widen the sources of money income. Incidentally, the extension worker can count on native hands nowadays to perform a number of Western skills, including carpentry, smithing, engineering, and radio work.

SPECIAL PROBLEMS. Several special problems that have arisen from modern contacts can receive only brief attention here, though they may well tend to monopolize the attention of extension workers in the areas concerned.

The first is that of tenancy and usury. These ancient institutions of Asia have been carried by Indian, Arab, and Chinese merchants and settlers into the more accessible parts of Malaysia, especially the lowland rice-growing areas. Where in other island areas the holders and users of land may at most give customary tribute or services to chiefs or leading families in recognition of their ceremonious authority over such property, here personal relations have long since hardened into landlord-tenant systems. Interest may be as high as several hundred per cent annually, and bond services or share cropping may be set at what by Western standards are fantastically extortionate levels. Tenancy and indebtedness are increasing, especially in crowded areas where population numbers are rising rapidly, and they are creating serious problems of poverty, exploitation, and social unrest. The Philippine census of 1939, for example, showed 60 per cent of Filipino families as owning no land. The John Doe of such countries is typically a tenant or share cropper caught in bondage and debt to a local or absentee landlord or an Asiatic moneylender, from whom he has little hope of extricating himself.

Colonial authorities have made some effort to meet these problems; but, in general, they have moved slowly, because the landlord class for the most part consists of the local elites who give the colonial administrators support and who tend to dominate any native representation in legislative bodies. Law and order is usually weighted on the side of the landlord, especially if the tenant tries to take the matter into his own hands by way of agrarian organizations or overt action. Laws have been placed on statute books curbing rates of interest, but these can easily be circumvented by the unscrupulous. Governments have also experimented with village or neighborhood banks and granaries, and have encouraged cooperative loan associations; in Java, even government pawnshops were established. Serious disturbances have already occurred in some of these depressed areas, and the extension worker faces difficult political as well as economic dimensions in his planning.

The population increase mentioned above amounts already in some places to overpopulation, and in many of these ocean- and mountain-bound regions population pressure threatens to become serious. So far, it has been felt most strongly in areas of two kinds: the crowded rice lands, especially around big city centers such as Batavia and Manila, and very small islands of limited resources. The Pacific islands have entered the "swarming period" of population increase that has touched country after country in recent decades. Birth rates, geared to traditional sex and social customs, continue high, but death rates are being progressively lowered through modern health measures and greater economic and social security. A few island areas are still undergoing depopulation, notably along the frontiers of the New Guinea region, because of temporarily exaggerated mortality caused by new diseases and other factors, but in general this formerly marked trend is now reversed.

Governments have tried to meet population pressure by helping to break in any new areas available for cultivation, by improving irrigation systems, encouraging more intensive production, fostering handicraft industries, and taking any other measures possible to widen the subsistence base in the home areas. Some have also experimented with schemes of resettlement and colonization. In theory it should be easy to meet the problem for a long time to come by allowing surplus numbers to spill over into the many relatively empty areas. To some degree this has been happening. Voluntary migration has taken hundreds of thousands of Filipinos, for example, to pioneer outer belts in the Philippines and even to the United States. Similarly, colonies of settlers from such tiny islands as the Tokelaus and Niue have taken form at the larger South Sea ports. Even so, the great majority of needy families are in no position to migrate because of want of capital, debts, and other obligations. Frontier areas are often disease ridden, as with malaria; local peoples (and spirits too) may be hostile; and, unless a community is formed that compares favorably with the friendly and secure ancestral settlement back home, such pioneering naturally has few attractions.

Government schemes have had to take these factors into account by establishing group colonies with adequate financial and welfare facilities. Inevitably they have proved costly, and so far have been minor in terms of the magnitude of the need. For example, in the Indies during the six years just prior to 1940, the Dutch resettled about 200,000 Javanese in the outer provinces, but in the same period Java's population increased by about 600,000 a year. An anthropologist visiting one of the small islands recently was asked in all seriousness by the elders whether the government would permit the age-old custom of infanticide once more. Such problems, though of course not unique to the islands, become especially sharp because of their definitely ocean-bound nature and because the islanders are rarely permitted to settle in countries elsewhere, even in the exceptional case that their usually tight ties with their ancestral communities become loosened enough for them to want to do so.

Some of the most serious problems facing the extension worker may also occur in relation to natives living in and around the urban centers and in the labor lines or barracks at plantations and mines. A small minority of natives in the cities and towns may have regular incomes as government or business employees, as landlords, and occasionally as successful professional or businessmen. The same applies to part-natives, the great majority of whom tend to congregate at such centers. But most natives, and many of mixed blood, are likely to have a much lower scale of living than the rural people. Besides those whose homes are in the area, the urban centers have formed a mecca for the ambitious, the discontented, and the very poor, and nowadays there is chronic unemployment. At most, the majority have to get along on casual work such as stevedoring, or by riding precariously on the economic coattails of luckier relatives. Native varieties of slums have sprung up, and health and welfare conditions may be very bad. In the war zones such centers have been bearing the brunt of trade dislocation, bomb damage, and other disorganizing conditions, so that their inhabitants are likely to be in greater need of relief and rehabilitation than any other groups.

Some jurisdictions still have systems of indentured or contract labor, by which native or other laborers are bound by a penal clause to stay on the jobs for which they have been recruited. Others have abolished this type of labor service or are restricting it, developing, instead, "free" labor systems. The days of "blackbirding" are gone; conditions of work, wages, housing, and other phases of employment are now regulated by law. Local officials supervise these matters, aided perhaps by special labor inspectors. Even so, the existing labor systems have been the subject of frequent investigations and offer a particularly difficult dimension from the viewpoint of welfare.

Health and nutrition are frequently poor among such laborers, the more so because official dietary prescriptions are usually antiquated in terms of modern standards. Women rarely accompany the men from their home villages, and, as there is usually little provision for constructive use of leisure, serious social problems have tended to emerge. The absence of a very large proportion of the younger men creates economic and social problems in the home communities. The extreme case of this kind has probably been the New Guinea Mandate, where almost one quarter of the adult males were serving as indentured laborers when war broke out.

The basic matter of nutrition in the islands deserves much more study than it has received so far. Limited information to date suggests that groups living primarily on a local subsistence basis tend to have adequate diets, though some native menus make for much better physique and teeth than others. (The amazing variety of native food items and the very different rhythms of food intake almost defy study by conventional "standards of living" schedules.) By contrast, groups living in the towns and labor lines are likely to have very inadequate diets. Polished rice, cheap canned meat and fish, sugar, and tea or coffee, often consumed with few fresh vegetables or fruit, open the way to beriberi, tuberculosis, and other diseases, the main incidence of which is in such areas. Infant diets have been made a matter of concern in some jurisdictions, and several government health departments have achieved notable success through setting up women's committees in native communities to supervise this and other aspects of baby care and family welfare. The extension worker may be able to encourage the development of small gardens in the urban centers and labor lines, and in other ways help to meet the problems of these special groups. But here, as in every type of activity with which the extension worker may be concerned, fundamental progress will be made only as the people themselves come to understand their problems and, with appropriate guidance, initiate the remedies.

FUTURE PROSPECTS. Over and above all these special conditions, and vitally influencing whatever solutions may be planned, looms the larger question of what economic future these island areas may have in the world that is now in the making.

As judged by recent trends, the outlook from the commercial viewpoint is by no means promising. Uncertain marketing opportunities, increasing competition from synthetics, and now the breaking in of great new areas of production to take the place of those temporarily lost to Japan, all point to straitened conditions in the postwar period. The question may justifiably be posed whether the welfare of the island peoples would not be better served by encouraging them to fall back as far as possible on subsistence economies rather than exposing themselves to the vagaries of world markets—becoming, so to speak, an island peasantry.

Nevertheless, important as it is to strengthen and stabilize the local economies, it must be recognized that the native peoples have definite and expanding wants which can be met only by entering to a corresponding extent into the larger commercial economy. Furthermore, the economic clock cannot be turned back on the cities and towns, the crowded areas, and the hopes of the educated youth. Extension programs must therefore try to meet as far as possible these legitimate needs by careful study and organization of local commercial production. Fortunately, natives usually command their own land and labor, and also have cooperative institutions that may afford a margin of advantage over nonnative individual enterprise. Among the leads that may be open are greater diversification of output, encouragement of any unique local crafts or other marketable skills and resources, and government assistance in handling and marketing of products. It seems certain that governing authorities, by way of extension programs or their equivalent, will have to take an increasingly active role in native economic life if these island peoples are to bridge without undue disorganization and hardship the gap between their traditional ways and the life of the modern civilized world.

Chapter 4 · CHARACTERISTICS OF PEASANT SOCIETIES · By Irwin T. Sanders

THE MAJORITY OF THE PEOPLE IN THE WORLD today live in folk societies. These "little people" of the world fill the teeming villages of the Orient, they form the bulk of the population in southern and eastern Europe, and they pass on from one generation to another the colorful costumes and customs of the Latin Americas. In North America evidences of the folk society are numerous: in the French-Canadian villages, among the peons of Mexico, and even in the more isolated areas of the Appalachian mountains.

At first glance it would seem ridiculous to lump together people of so many languages, of so many creeds and colors. However, underneath their apparent divergencies are many ways in which they are much alike. This chapter briefly describes some of the characteristics which are most important in agricultural extension work. It is not intended to be a rigorous systematic treatment of the pertinent anthropological and sociological literature. The purpose is to help the person trained in the Western World (or in institutions modeled after those in the Western World) to gain an insight into the folk mentality and the peasant way of life.

FOLK SOCIETY A SURVIVAL. One of the most important features of the folk society is that it has stood the test of time. For centuries it has been a way of life to countless millions, carrying with it the authority of past successes. But now the various folk of the world are finding that they must adjust in part to a twentieth-century mode of living which is as yet virtually untried. It is rather an adventure of man's spirit into the unknown, an adventure which invariably involves a casting away from the moorings that formerly gave the greatest feeling of security. No wonder, then, that peasants look with suspicion and reservation at the new world dawning about them. The folk society represents the tried and true, a survival. The new ways have yet to prove themselves.

Wallis in speaking of the folk society says "Here in custom, language, belief, handiwork, and technology one finds the record of bygone days, a section of the past persisting in the present, like some geological stratum laid down millennia ago which thrusts upward into recent surface soil a weather-worn testimony of old accretions." [1]

Thus one must appreciate the historical continuity of a folk society if one is to understand its nature. It was centuries in the building and it still has a strength that hurried Westerners often fail to grasp. Although breaking down in places, like a windmill perhaps rusting and squeaking, it can still prove more than a match for modern Don Quixotes battling with more selfrighteous enthusiasm than social wisdom.

ROOTED IN THE SOIL. Members of a folk society are of the earth, earthy. They vary from climate to climate and country to country in the fervor with which they till the soil, but basic to their way of life is the security which agriculture affords. When they choose to work, providing the powers that be are propitious, they know they will reap the fruit of their labors.

In most folk societies much is made of fertility rites, which emphasize fertility of soil as well as fertility of women. Where the cultivation of crops is laborious and where the pressure of population upon food supply is great land hunger grows and a society becomes even more deeply rooted in the soil. This is not to say that labor per se is necessarily glorified, but rather the object of the labor—the soil. Few peasants are sophisticated enough to think of labor as one of the factors of production. Labor is a constant which is taken for granted, whereas the quality of the soil and the caprices of the weather are the variables which interest the people.

Life follows the cycle of the seasons. Because of this, the year has an accentuated variety. The tedium of monotony is periodically broken. The tasks of spring differ from those of fall; the leisure of winter is in striking contrast to the demands of summer.

Thus all of life is strongly colored by the man–land relationship in a folk society. Whether one owns the land oneself, or works for years for those who do, one develops a sense of proprietorship, at times mystical and at times strikingly realistic. Under the Czars the Russian serfs had a saying which combined both mysticism and realism and showed that the peasants felt they had the best of the bargain: *My*

[1] Wilson D. Wallis, "Folk and Cultural Sociology; Methods of Analysis and Utilization of Results," *Publications of the American Sociological Society*, XXVII, No. 2 (May, 1933), 77.

vashi, zemlia nasha. "We belong to you but the land belongs to us."

USUALLY FAMILISTIC. Ancestor worship in China may seem to have little in common with a feud in the Kentucky mountains, but in reality both are evidences of the same fundamental belief that the family (with little or no help from the government, from business, from the church) is capable of carrying on the important affairs of life. In a folk society the family tends to occupy two spheres of dominance: over other institutions and over the individual.

The Family Is the Central Institution. In most folk societies the family is the chief economic unit. It produces not only the food but it also makes the soap, the homespun cloth, and the crude furniture: the family frequently constructs the home. In the person of some elder relative acting as midwife or doctor it brings infants into the world and treats the sick with mysteriously concocted remedies. In other words, the economy of the home is largely self-contained. Provision is usually made, however, for specialization along lines which serve the interests of the group as a whole. A wagonmaker, a tanner, a tinsmith, a master builder, a wood carver—these and many more illustrate the contribution of artisans to the physical welfare of the folk society. As the money economy began its encroachment, specialists in money-lending and merchandising became a part of the picture. Despite their coming, the family group was still considered fundamental and these artisans a mere matter of convenience.

Nor was formalized government of much importance in years gone by. Elders, representing large family groupings, conducted the affairs of the community in what they considered to be the best interests of all. In familistic societies throughout the world the people still look with resentment upon the imposition of outside officials who try to enforce laws and local regulations alien to the folkway of rule by older family representatives. Local leaders who are related by blood to the villagers have much more influence in the formation of public opinion than do the officials acting in the interest of a centralized government which the people feel they had no part in creating.

Religion, too, has its place in a familistic society but proves a bulwark rather than a competitor of the family system. Quite often the religious leader does little more than officiate when family crises arrive. In areas where Christianity has been accepted by large numbers of people, the Roman Catholic Church has tended to preserve the folk society while

Protestant influences have tended to tear it down, moving it in the direction taken by Euro-American society in general. Islam and the other religions of the East are chiefly protectors of the status quo in which loyalty to one's family seldom conflicts with loyalty to one's religious duty.

Educational agencies and organized sources of recreation are little developed in a folk society. Much training is given in the home, and recreation is supervised by the family or by community opinion in general. Where school systems have been established by outside authorities—governmental or religious—they have tended to neglect the values of a folk society, emphasizing the values of Western society in which the school has an important role.

The Family Comes First, the Individual Second. The individualism of the West contrasts strikingly with the familism of the folk society. In the latter, customary controls still govern family relationships, although the conjugal family, made up of parents and children, is displacing the consanguineal family, comprising, in addition, many relatives of one of the parents. The kinship group traditionally guides the selection of mates by the young, and those earning money outside the home must turn these wages over to the family treasury; in turn, individuals wishing vocational training or education can count upon the support of their relatives, providing their plans follow the family decision. There is also a well-defined division of labor which is based not only on one's sex and age but also upon one's status in the family. Traditionally, the father maintains his dignity by being distant from his children; the mother expects a life of constant toil and frequent childbearing; the children expect little consideration, and they early learn to avoid those things which bring down swift punishment upon their heads.

Linked to this subservience to family roles is the fact that one's status in the community largely depends upon the status of one's family. An erring individual reflects upon the family, because in a folk society there has been no attempt to place the burden of moral training upon the school or the religious organization. It remains a family responsibility. Thus, in the community at large, who you are (what family name you bear) is more important than what you are as an individual. Nevertheless, relationships remain personal, in contrast to the impersonality of the West, where what you have is frequently more important than what you are, or even the family to which you belong.

Conservatism of Folk Society. The conservatism of peasants is rooted in their social values. Because their way of life provides what they consider their most important needs they are satisfied with it. To be sure, social inertia is present too. It is far more comfortable to cling to the old than to embrace the new. Furthermore, in a folk society individuals who seek to adopt a new practice frequently face the ridicule or even the hostility of their fellow peasants, primarily because of what has been called the strain toward consistency, or the tendency to pull all exceptional individuals down to the level of the average. Individual initiative, which is an important element in innovation, is discouraged by the less ambitious. If a person finds such an environment too restrictive he may migrate to the city and become a part of the urban world, with its different set of values and its impatience with the peasant way of doing things.

The folk society still lacks many of the agencies of communication which make modern life so dynamic. Automobiles and good roads, which encourage extreme mobility, are not yet a part of the folk culture; neither are private radio sets, or family subscriptions to daily papers or memberships in monthly book distributing societies. Communication is oral; the printed page is of little importance even though the literacy rate may be high among certain groups of peasants.

Whether it be a cause of his conservatism or its result, the member of a folk society lives in the present. Of course, the dead hand of the past controls his movements to the extent that it points the way through custom to one activity after another, such as putting blue beads around a donkey's neck or eating a prescribed dish at a particular feast. But in any event there is little apparent concern with the future. The peasant, unless influenced by modern commercialism cannot conceive of investing a sum of money now in order to get back twice as much years later. This is why it is difficult to put across a forestry program. Individuals see nothing wrong in letting their hungry goats strip the young trees of leaves, for they are not accustomed to looking ahead ten years. They take each day as it comes.

This trait explains as well why members of the folk society will traditionally practice soil conservation where they can see the results in this year's or next year's crops but will do little to build up the soil for benefits to be derived at some more distant time. It is only as one begins to translate courses of action into monetary terms that one begins to sacrifice for the years to come. A barter economy therefore is one centered in the present; only in a money economy can one deal in "futures" whether on the grain exchange or in the speculation of land. By and large, the folk society is a barter economy, where "futures" come low.

SPATIAL AND MENTAL ISOLATION. Part and parcel of this conservatism is the peasant's isolation, spatially and mentally. Roads, for example, exist primarily for use between village and field or between village and market. Good roads in and of themselves rate low in the scale of values. So does going somewhere just for the sake of going. If a pilgrimage is customary then the person will take a pilgrimage. He does not hanker after the strange sights and experiences of a traveler and, in fact, he usually finds these disconcerting, unenjoyable, and even terrifying. But, upon returning to his native haunts, he does bask in the limelight of being called a Hadji, or by some other title which shows that he has followed a time-honored tradition.

In some societies the people believe that the soul of a deceased person must retrace within a matter of days all the earthly journeys of a lifetime. This proves a difficult chore for the spirit of a much-traveled wanderer. While it reflects the social pressure toward permanence of residence, it does not mean that some individual who has been to a distant place loses status upon his return. His tales enliven the neighborhood gatherings and, since he usually casts aspersions upon the food, dress, and manners of people abroad, he strengthens the ethnocentric beliefs of his fellows in the superiority of their own way of life. He does lose status, however, if he compares his own folk or society unfavorably with the "outside world," for to that extent he becomes a nonconformist, a disloyal member of the local group.

There is a marked cleavage between the village, the perpetuator of traditional folkways, and the city, the culture-carrier of the scientific age. For one thing, many city dwellers of peasant origin who have accepted the newer social values of the West have the zeal of new converts in ridiculing the life from which they have come. Although familiar with what the peasants do and why they behave as they do, these converts develop a bias which makes them highly unreliable as social interpreters. Government bureaus in many capitals are filled with these erstwhile peasants zealously struggling for urban status. Then, again, those whose families have been city dwellers for decades have lost touch with the rural people; they regard them as objects of idle curiosity or as fit subjects for exploitation.

Therefore, where the folk society still maintains a pristine vigor, the peasants build up a psychological barrier between themselves and the city dweller which frequency of urban contacts may actually strengthen. The peasants, in other words, do not succumb to the advertising patter which implies that because a thing is new and up-to-date, it is necessarily superior. He evaluates a new article or a new proposal in terms of the accustomed or conventional rather than on the basis of its novelty.

COMPARATIVELY LOW STANDARD OF LIVING. If one should seek an illustration of the changes wrought in Western society by the machine in the past one hundred years let him look first to a folk society. Using this as a starting point modern man can measure his material advancement. Because agrarian societies have an abundance of labor and think seldom in terms of profits, they have little incentive to mechanize their practices in home and field. In many ways they consider the machine a threat and find numerous pretexts for resisting its introduction. They praise what has been made by hand and hunt for flaws in machine-made products.

They produce, where agricultural conditions and size of farms permit, enough for home consumption, with a small surplus to exchange at the market for absolute necessities (such as salt and nails). Living conditions are crude and unsanitary; death rates are high. But this waste of health and life is a part of the fatalistic creed to which most of them subscribe. They have long ago shifted the responsibility to the supernatural and thus are able to adjust to the numerous bereavements that befall them. What the Westerner most frequently associates with a low level of living—unscreened windows, unsanitary handling of meat, improper disposal of sewage, impure and insufficient water, overcrowded houses, raw-boned animals, improper diet—seem relatively unimportant to a peasant. He is more concerned about having sufficient quantities of the kinds of food he likes, having a place to sleep, protection from the weather, and the wherewithal to observe the year's festivities in the recognized fashion. He wants to have enough fuel to see him through the winter, enough provender to keep his animals alive. In some societies, for example in India, many peasants follow an economics of scarcity, calculating how little they will need to grow for the year ahead and concentrating on producing this amount.

Most efforts at extension will quite properly be directed in peasant countries at eliminating hazards to health and improving agricultural production. What the extension worker may not at first realize is the reason why people need to be prodded into accepting things which are self-evident and commonplace to one familiar with them—for example, the germ theory of disease. Of course, the peasants want to live a long time. Old people are honored, but a long life span is a gift of God. Presumptuous is the man who would assume the prerogative of the Eternal, and as for increased production, peasants know only too well that when crops are bountiful prices are low. Or when prices are high the things they want to buy are even higher. So the average peasant plans his yearly activity with his immediate needs in mind, foregoes the luxury of spending in his imagination large sums realized from next year's crops. He counts his pennies closely, he at times becomes penurious, but he does keep going surprisingly well from year to year.

AN INTEGRATED SYSTEM. Robert Redfield has effectively stressed the unity found among the various phases of the folk society. In fact, he points out that the ways the group meets its problems "constitute a coherent and self-consistent system." He goes on to point out:

Thus it is not enough to say that in the folk society conventional behavior is strongly patterned; we must also say that those patterns are interrelated, in thought and action, so that one tends to evoke others, and to be consistent with others. We may add to this that the more remote ends of living are taken as given. The folk society exists not so much in the exchange of useful functions as in common understandings as to the ends given. In the trite phrase the folk society has a "design for living." [2]

Because of this design, large numbers of people who differ from one another in many minor patterns nevertheless feel a sense of belonging to a peasant class. True enough, most of their association is in the primary groups of family and neighborhood or in the larger unit of the village community. But where one goes as a son-in-law or a daughter-in-law to a village many miles away one can readily adjust to such differences in plowing, breadmaking or embroidering a shirt. This is

[2] Robert Redfield, "Rural Sociology and the Folk Society," *Rural Sociology*, March, 1943, p. 70.

because the social values learned in the early home hold good for the new locale. This awareness of or identification with a peasant class is often vaguely defined among the members of a folk society. Its latent presence, however, is convincingly shown whenever there is an issue between peasants and nonpeasants. The villager will usually stick with other villagers even though they are strangers to him. One can thus accurately speak of the peasant mass and partially understand its staying power in the face of depressions or changing governments.

This unity of the folk society has another important implication. Changes which vitally affect one phase of life have their repercussions all along the line. Nowhere today does the folk society described here exist as a pure type. Everywhere it has taken on some of the characteristics of the Western society. It operates more and more on a money economy, its members are increasingly being drawn into the orbits of centralized governments, commercial interests are making the peasants trinket and gadget conscious, young people grow restive under the age-old conventions, especially when cities present attractive alternatives. These changes would not necessarily destroy the folk society, which still retains much assimilative power, but their effects deserve careful weighing in terms of each society. Those who seek to change bear a responsibility that cannot lightly be cast aside, for each innovation launches a train of events which in the long run may produce results not originally desired. This does not mean that the only policy is a "hands off policy," but it does call for careful study of the society one wishes to change. Change there will be. Much will be impersonal and seemingly beyond control. But the consciously planned program of reform, uplift, or regeneration must be thought out and executed against the background of the social fabric, which in a folk society comes all of a piece.

Chapter 5 · PROMOTING COOPERATIVE AGRICULTURAL EXTENSION SERVICE IN CHINA
By Hsin-Pao Yang

NATURE AND IMPORTANCE OF CHINESE AGRICULTURE. China is probably the oldest and largest agricultural country in the world. The total area in farms has been roughly estimated at 232,000,000 acres, which supports as many as 450,000,000 people. There are approximately 60,000,000 farm households in the country out of a total aggregate of 80,000,000.

Chinese agriculture is, first of all, the source of food for about one fifth of humanity. The most important crops of China are rice and wheat, which are used as the staple foods. Cotton is produced for clothing and other textile purposes. Other crops, amounting to about one per cent or more of the total crop production, are, in order of importance, millet, soybeans, kaoliang, barley, corn, sweet potatoes, rapeseed, broad beans, peanuts, green beans, and field peas. Mulberry trees, tea, tobacco, and many types of citrus fruits, vegetables, and nuts are also important in the Chinese rural economy. The principal livestock are oxen, water buffaloes, and hogs. About three fourths of the livestock are used for draft purposes and only one fourth is raised for meat. Sheep, horses, mules, and donkeys constitute a rather small percentage of the total animals raised, and are regarded as of little economic importance. Poultry raising, though common in rural China, accounts for an insignificant part of the national income.

Chinese farming still lingers in the stage of "hoe culture" in which it takes four peasants to produce enough food in normal times to feed their own and one extra family. In other words, about 80 per cent of the Chinese people must engage in farming in order to feed the whole population. This rate of productivity is extremely low compared with that of the United States, where, at present, nineteen American farmers are producing enough food to feed themselves and as well as sixty-six city people besides. On the basis of production per man-equivalent, according to another estimate, one Chinese farmer working a full year produces only 1,400 kilograms, or approximately 3,000 pounds, as compared with 20,000 kilograms, or approximately 44,000 pounds, produced by one American farmer. One American farmer is producing fourteen times the amount claimed by his fellow worker in China, with far less hardship and risk.

The size of land holdings in rural China has been variously estimated by different authorities. One study indicates that about 16 per cent of the Chinese peasants own 30 to 50 *mow* (one *mow* equals one sixth of an acre), 24 per cent own about 10 to 30 *mow*, and about 44 per cent own one to 10 *mow*. For the nation as a whole, the average size of a Chinese farm is estimated at about 20 to 30 *mow* (3 to 5 acres), which not infrequently are scattered in small strips.

The low rate of productivity and the small holdings make Chinese agriculture precarious and uneconomical. The farmers are not producing enough food for China's ever-growing population. It was reported that in the prewar years of 1933–1937, China had an average net import of 50,772,000 bushels of rice and 15,823,000 bushels of wheat. These imports were consumed largely by the ports and cities near the coast.

Handicapped by the insufficiency of food, which is further restricted by transportation difficulties, the Chinese diet is much more bulky, lower in fat, and less digestible than the mixed diet of Westerners. It is a deplorable fact that the Chinese people are suffering from malnutrition, a problem needing further careful study and analysis if the national welfare is to be improved.

CULTURAL STABILITY. The Chinese farmers for the most part live in villages. These communities reflect the comparatively simple life of primary groups in close touch with the soil. The village is a social cell. It still retains many characteristics handed down from ancient times. In the village, people are largely linked together through blood relationships. In this isolated social cell, people manage to earn a living by cultivating the fields inherited from their forebears. They make their own homes among the earth-built cottages of their kinsmen. They go to the same markets, buying and selling in the same way. They use practically the same articles and wear the same apparel. They marry and are given in marriage according to the same customs and following the same traditions.

In the village, people are inclined to follow the same patterns of social behavior with respect to rearing and educating their children, observing and believing in certain religious ideas and practices, and enjoying—though rarely—a few amenities of life.

In brief, the Chinese village is a family village. It still retains remnants of an early familistic stage. The bond of family life and the bonds of village life are one and the same. It is in the village that most people manage to earn their living, make their homes, raise and train their children, enjoy their leisure hours and worship their gods according to the dictates of their conscience. It is here, as thirty centuries ago, most of the villagers—free, independent and self-reliant people—would sing:

> When the sun rises I work,
> When the sun sets I rest.
> I dig the well to drink,
> I plow the field to eat.
> What has the emperor to do with me?

Agriculture, being a form of social heritage, is a way of life to the Chinese people.[1] They live in agriculture, move within agriculture, and from agriculture they derive their principal sustenance. It may be said that the Chinese traditionally have possessed agriculture just as much and as tenaciously as agriculture has possessed them since time immemorial. The age-long dependence of the Chinese people on Mother Earth produced certain patterns of social relationships and other inherent traits which characterize the plain, simple lives of millions of the Chinese peasants. Countless Chinese peasants have thus been born into and lived through a typically agrarian culture. These accumulated practices are their sacred heritage.

Normally speaking, people are resistant to change; the Chinese is no exception. It has been alleged that during the past half century China has been rather slow to adopt modern ideas and practices. True as this might seem, it needs to be stressed, however, that change and readjustment could not and would not be accelerated until the people felt ready for it.

"Changing agricultural practices," said Dr. M. L. Wilson, Director of the United States Federal Extension Office, "is changing a way of life."[2] It is bound to be slow and is always fraught with vehement

[1] According to L. H. Bailey, farming is the primitive and underlying business of mankind. Agriculture is not a technical profession or merely an industry, but a civilization. It is concerned not only with production of materials, but with the distribution and selling of them and with the making of homes directly on the land that produces the material. *The Country-Life Movement in the United States* (New York, 1911), pp. 14 and 63–64.
[2] M. L. Wilson, "Thomas Jefferson—Farmer," *Proceedings of the American Philosophical Society*, LXXXVIII, No. 3 (1943), 221.

opposition on the part of the farmers. However, ideas favoring change and innovation are fermenting in China. Among others, one school of thought tends to stress the belief that China must first modernize her agricultural systems and educate her peasant folk before any large scale of industrialization can be efficiently launched. To achieve these two broad objectives, many measures are now being proposed and considered. In the following pages we shall discuss briefly the desirability and practicability of promoting the cooperative agricultural extension service in China.

ATTEMPTS AT AGRICULTURAL EXTENSION. Against the background of a comparatively stable culture ripped suddenly open by the impact of Western civilization and technology, a few initial steps have been taken to improve and develop the Chinese agriculture and to enrich the lives of the vast number of Chinese peasants. Agricultural extension service, according to Dr. P. W. Tsou,[3] is a new phase in the history of Chinese agriculture. The system as such is still in its embryonic stage. Much credit must be given to a few outstanding individuals and private agencies who first introduced and applied this system to China. According to Dr. Tsou again, the first attempt was made by the National Committee of the Chinese Y.M.C.A. Around the year 1915, under the able leadership of the late Dr. David Yui, it sponsored a popular lecture tour on reforestation. This program was carried out largely in big cities for nearly two years and was enthusiastically accepted by the audiences.

In 1918, Dr. John H. Reisner, then Dean of the Agricultural College of the University of Nanking, introduced the extension system to improve cotton production in China. One cotton specialist of the United States Department of Agriculture was invited to direct this work. A few varieties of cotton seeds were brought over to China and were grown in various localities. As a result of this preliminary, experimental work, the knowledge of modern agricultural science was extended to the Chinese farmers.

The scattered efforts of individuals and private agencies in the field of agricultural developments prophetically indicated the possibility of employing agricultural extension as an instrument of rural educa-

[3] Resident Representative of the Chinese Ministry of Agriculture and Forestry, Washington, D.C. His "Agricultural Extension in China" is a statement submitted to the conference held in Washington, September 19-22, 1944, to Outline the Contribution of Extension Methods and Techniques toward the Rehabilitation of War-torn Countries.

tion to bridge the gap between agricultural scientists and the practical farmers. Space will not permit a more detailed review of the good works carried on by different pioneers. However, a few outstanding examples may be cited in passing.

Perhaps the first to be mentioned is the Mass Education Movement. The Ting-Hsien Experiment represents the most unique feature of rural education and rural reconstruction ever tried with considerable success in China. The movement was initiated by Dr. Y. C. James Yen as a campaign for eliminating illiteracy. Its first endeavor was to simplify the process of teaching the Chinese peasants to read and to write. Its well known "Thousand-Character" texts or readers gained nation-wide reputation. Ting-Hsien also made some constructive demonstrations in what was regarded as a fourfold fundamental educational program for the rural communities. The first program—a distinctive contribution—was literacy education. Its second program was concerned with agricultural and economic reconstruction; the third dealt with rural health, and the last, with citizenship education.

The second private group to gain distinction in rural extension was the Li-Chuan Christian Rural Service Union. This experiment represented another approach to tackling rural problems by the combined efforts of the Christian Church and the local government.

A number of universities and colleges also made some early worthwhile contributions in the field of extension service. Among these, the College of Agriculture of the National Southeastern University was the first to undertake large-scale improvement work on cotton. According to Dr. Tsou's account, it was from this institution that a number of experimental stations were established in different cotton producing provinces.

Next in order is the Agricultural College of the University of Nanking. As early as 1924, a division of extension was established in this Church-related institution. Because of its well-trained personnel and splendid equipment, this agricultural college stands today unrivaled in the field of agricultural training, experimentation, and extension work in China.

Lingnan University of Canton Province was another private educational institution which did outstanding research in seed selection, fruit culture, soil analysis, insect control, crop rotation, land-use practices (in sub-tropical regions) and home economics.

Fukien Christian University in Foochow, Fukien Province, pioneered in 1934 the organization of a rural service center. The work of this institution proved that the best way to serve rural communities is to work with the rural people, meeting them where they are, rejoicing with them in prosperity and sharing their burdens in time of affliction and adversity.

Besides these private colleges, some government-supported institutions also did notable work in rural extension service. To mention only a few outstanding ones: the Teachers College of Wusi, Kiangsu Province, the National Peiping University, the Tsou Ping Experiment in Shangtung Province.

Besides the private agencies and the educational institutions, the work done by the Chinese Government needs to be mentioned because of its potential and actual influence on the improvement and development of Chinese agriculture.

Unfortunately, little can be said of the work done under the old Peking regime. According to Dr. Tsou, an extension office of agriculture was created in 1915 by the Ministry of Agriculture. It was situated in an old temple outside the city of Peiping, where small plots near the temple were set up to raise a few foreign varieties of field crops as a demonstration. Attempts were made to distribute the seeds to near-by farmers. This office was abolished in 1917, but was, however, the first central office of agricultural extension established by the Chinese Government.

Since its establishment in 1927, the Chinese National Government has spared no efforts to promote agricultural production. Promising results have been obtained in food and cotton crop improvements. In 1929, a National Committee of Agricultural Extension Service was organized. Governmental agencies which took special responsibility for different phases of agricultural developments were the National Land Administration, the National Water Conservation Commission, and the Farmers' Bank of China. Among the most recent additions have been the Ministry of Social Welfare, the Ministry of Food, and the Ministry of Agriculture and Forestry.

The Ministry of Agriculture and Forestry is the central agency responsible for the planning and administration of the nation's agricultural affairs. Besides its regular administrative machinery operating on the national level, there are three important bureaus for scientific research: the National Agricultural Research Bureau, the National

Animal Husbandry Research Bureau, and the National Forestry Research Bureau.

In 1943, the Chinese Agricultural Association drafted a preliminary outline for postwar agricultural reconstruction in China. This draft, after a period of intensive study and examination by professional people, was released to the public for further discussion. Among other proposals, the draft recommended the reorganization of the Ministry of Agriculture and Forestry by creating nine Bureaus, one of which is a Central Extension Office. The Central Extension Office is to serve as the general headquarters of all agricultural extension activities in the country. Four divisions operate within it. The first deals with rural cooperatives; its official responsibilities will cover such items as agricultural production, processing, marketing, distribution and farm insurance. The second division is to deal with land-use planning and development, farm implements, and irrigation. The third takes care of informational activities, while the fourth will have charge of field coordination and assistance. Attached to this last division will be a number of regional offices each of which will maintain a traveling demonstration team. Each demonstration team will be composed of a number of subject-matter specialists who will keep in close contact with different extension offices throughout the country.

Under the Central Extension Office there will be several Provincial Extension Offices. From the Provincial Office the line spreads downward to the District (county) Extension Office, and finally reaches the local cooperating bodies.

EXTENSION PROGRAMS TO DATE. The foregoing brief accounts only serve to record a few early attempts in extension service in China. Forty years have gone by since agriculture was first admitted to the regular school curriculum. These past records, indeed, can hardly impress the reader, for looking across the whole field, one cannot fail to realize the immensity of the job and the meager beginnings represented in the past efforts. The lack of success may be explained in terms of obvious handicaps. In the first place, the low income of the Chinese farmers, the conservatism of the rural communities, and the prevalence of illiteracy prohibited any ambitious schemes of extension work during the initial period. Other obstacles were the lack of facilities of communication and transportation, lack of a mature political body to exercise control and to achieve efficient administration, lack of trained and competent leadership, lack of technological equipment and of the supervision of experts and specialists, separation of planners and administrators from the living realities of the farm people, arrogance and petty snobbishness of a few (fortunately few) so-called modern trained experts and specialists, universal window-dressing of politicians whose ulterior motives overshadow their faith in agriculture and in the rural people themselves, and finally, continuous social unrest and tumult resulting from internal conflicts and external interferences.

Observing more closely the road through which extension work has just passed, we find that the Chinese have made a few mistakes which need to be corrected. In the first place, many of the private agencies and institutions have been operating with insufficient funds. Many of the pioneering people were motivated by a charitable or missionary purpose. In most instances, they launched the movement with heated enthusiasm, thinking little about the inadequacy of their financial strength. Once started, however, they found the job too big for them. When their funds were exhausted, the fire of their original zeal flickered out too. Quite a few of the early experiments in extension activities suffered this fate, leaving many noble souls in utter disillusionment and disappointment.

Secondly, among the public or government agencies, the most glaring weakness was a lack of real or serious purpose which spelled failure even before the project was started. Many of the government programs for agriculture were not formulated with any abiding faith in agriculture nor with any genuine interest in serving the rural people. Sometimes these governmental measures were proposed only for political advantage. Once the ulterior purpose was served, the proposals were dropped. If a proposal survived the preliminary paper-work stage and evolved into a thorough-going organization, its "busy" Commissioner devoted much of his time to arranging and displaying some brilliantly painted signs over the gate entrance. He also would provide himself with an abundant supply of letterheads and some printed or mimeographed materials which contained elaborate organizational charts and graphic presentation of the rulings governing the functions of different offices. All the preliminaries and paraphernalia were planned with earnestness and usually worked out to the minutest details. Having done these, the Commissioner considered his assignment finished, attempting no more active program than was expediently necessary and leaving the organization to drift with political winds.

Another common mistake found among private and public agencies is the tendency to overemphasize imported materials and neglect the use of local products. In a few isolated agricultural experimental farms, not infrequently one would find plots of corn, tomatoes, strawberries, radishes, spinach, and asparagus, carefully marked with foreign names and origins and methodically protected by the station masters. Growing in the adjoining fields, however, one saw patch after patch of local beans, mustard, turnips, onions, garlic, peas, and Chinese cabbage, where insects were having a field day. But the station masters did nothing. They explained apologetically that they knew very little about the local produce, and so could give little help in its cultivation. These same masters in their annual reports, however, would complain about the stubbornness of the peasants in their failure to adopt those new practices which they tried to introduce.

As a result of all this, another serious blunder was made by a few institutions which foolishly spent a lot of money in buying foreign farming machines, not for practical use but only for exhibition. One amusing instance was found in a certain agricultural institute in South China. The first consignment, shipped to the campus immediately after the school was opened, was a vast purchase of German-made tractors, threshing machines, and other motor-driven implements. These scientific instruments were housed in a specially constructed exhibition hall where crowds gathered around to see them once a year during the observance of Founder's Day. Interestingly enough, right on the Institute farm, where sugar cane and sweet potatoes were being grown experimentally, the coolies were still using the same type of hoes and cultivating with the same type of wooden plows which their forebears made a few thousand years ago. Here again the exhibitionists would raise their eyebrows to question why these "ignorant" farmers could not appreciate the supreme value of the modern machines.

Lastly, we have seen a few Chinese scientists and experts who have been well trained in theories and principles. They were indeed unusual teachers in classrooms and maybe very entertaining in the laboratories, but strange as it may seem some of them would hesitate to go out in the field and soil their hands. A college professor of forestry was reluctant to go out on a rainy Arbor Day to teach the near-by villagers the proper way of planting trees. His reluctance would be better understood if he were a college physician instead of a strong protagonist of reforestation. To the simple-minded peasants who had walked three to five miles to observe the much advertised demonstration the effect of the weather on the expert was a gross disappointment.

EXTENSION LOOKING FORWARD. Extension work is based, as it should be, on the practical principle of learning by doing. Farmers should be encouraged to learn new things by doing and by direct participation. To make extension work effective in China, emphasis upon theories and abstract principles must be accompanied by a common-sense approach.

Teaching scientific farming to those who farm by tradition is no easy task. A decade ago a sensation was created among the villagers when a college professor took off his shoes and got himself dirty in a rice seeding plot where he demonstrated the proper way of transplanting. He put his teaching across because he followed the most natural way of working with the people. In the following pages are suggested similar approaches which may be helpful to those who wish to serve the farmers best with the least resistance. A few failures are cited by way of illustration. Such cases fortunately are becoming more rare and, as agricultural science takes its root in China, these mistakes are bound to disappear.

Extension Work Is a Grass-Root Operation. Rural reconstruction should start at home. The best way to serve the rural people is to meet the interests that are close to their own hearths, fields, and graveyards. It is distressing to find a few returned students who can relate vividly how cotton is raised in Mississippi, corn cultivated in Iowa, wheat harvested by combines in Kansas, but are unable to help the hard-struggling Chinese farmers in their potato patches or in their rice paddies. Here is a plain fact which the agricultural scientists and experts might fail to realize; namely, that any endeavor designed to serve the interests of the farmers should be bound up with and determined by the needs and desires of local communities.

Effective Extension Work Relies on Local Leadership. The importance of local leadership cannot be overemphasized. Experienced extension workers rely on it as an agent to get things done. This simple fact was most forcefully brought to the writer's attention when, as Director of the Provincial Institute of Mass Education of Fukien Province, he sent three trainees to establish folk schools in a certain community. Because no headway had been made within three months' time, it was necessary for the Director to visit the community. Upon his arrival he discovered that his three representatives had ignored "old Chang," who was the "head-eye" man of the community. A social get-together was immediately arranged to which "old Chang" was invited and accorded due respect. Within three weeks three folk schools were opened. "Old Cháng" gave his blessing and his full support once his position as the "boss" of the community was recognized. Perhaps it is no exaggeration to assume that in most folk societies and in a few geographically isolated communities few outsiders could possibly succeed in getting things done without the approval of the local leaders. To ignore them is to put sand in the machine, to use them is like lubricating the motor with oil.

Local Materials Should Be Used. There is a lesson for us in the story of buying foreign-made machines for exhibition. Imported materials may at first appear attractive but when locally applied may meet with little success. In a suburb of a city in South China, a commercial agent tried to introduce a certain brand of German-made fertilizer for use in cauliflower production. The result was definitely disappointing. It was financially beyond the reach of the average farmer to obtain this expensive and supposedly superior chemical and, what is more, neither the farmers nor the commercial agent knew exactly how the soil would respond to it. Here the agronomist in a near-by college came to the farmers' rescue. He began to analyze their soils and advised them to process their barnyard manures. Within a short period noticeable improvement was found in the cauliflower crop.

As this simple story shows, agricultural education begins at home. Extension workers should look right at their doorsteps where useful materials await exploration and exploitation. In the kitchens, in the woodpiles, in the cowsheds, along the streams and on the hillsides, there are potential resources which extension workers can employ to the genuine satisfaction both of themselves and of the local people. Chinese extension workers can ill afford to advise farmers to use T.V.A. phosphate while forgetting that the average farmer, like Wang Tsung, has only three acres to operate. Imported chemicals might not bite Wang's soil too badly, but they certainly could ruin his pocketbook, if he had one at all.

Following Psychological Pathways. Farmers, generally speaking, are encumbered with heavy psychological baggage from the past. These accumulated odds and ends have in previous times provided some genuine satisfaction to the people. To throw these old practices and old things away could easily arouse deep-seated resentment. A public health worker was almost expelled from a community because of her open criticism of the way country folks fed their nursing mothers immediately after babies were born. All mothers ate fried chicken daily and avoided taking green vegetables and drinking cold water. The nurse thought that the diet was wrong and maybe harmful. But the womenfolk, as well as the menfolk, all regarded the eating of such food a rare privilege granted only to those who gave birth to babies, especially boys. To take fried chicken away from the nursing mothers was to create personal disappointment, community conflicts and disorder; to question and criticize this practice was to produce feelings of distrust, hostility, insecurity, and social frustration on the part of the local people.

Again, this simple story tells a plain truth; namely, that in the village, people live by well-established behavior patterns. It is better for the extension worker to work with, not against, the existing customs and habits of the people. Both customs and habits are dynamic stabilizers of community life. To ignore people's traditions and customs or, worse, to work against them is to create suspicion and intensify the feelings of resistance.

Play Safe with Farmers. "To abide with time, play safe with farmers" is a colloquial expression in China. Farmers in a folk society need to follow the slow yet steady ways of doing things and find it necessary to play safe with their production and cultivation. When six or more persons have to live on a three-and-a-half-acre farm, the possibility of experimenting with new crops and with new ways of cultivation is very slim. A personal friend of the writer once tried to move ahead too fast in a poultry improvement project. This poultry specialist asked farmers to exchange their eggs for his quality eggs in order to hatch better chicks for their farms. People did not want to cooperate. Upon investigation the poultryman found that the farmers disposed of the "foreign" eggs, instead of hatching them. Their reason for such action was that they could not afford to furnish all the necessary equipment for the "foreign" chicks, as recommended by the poultry specialist. They were actually afraid to use the high quality egg which involved many financial risks and some inconveniences. It took quite a few years before the poultry specialist did succeed in educating a few farmers and inducing them to use his quality eggs.

"Slow and sure" is, therefore, a very practical and expedient slogan for working with farmers. There is always the time element to consider. The fervor of a reformer or the zeal of a crusading prophet most likely will not help but rather hinder an extension program. It pays well for extension workers not to do everything at once and too quickly. They should not force people to run before they learn to walk. Extension workers should also realize that human nature, if it changes at all, does so with a "geological slowness," in which education rather than force is the guiding principle.

Avoid Imposing an Overhead Structure. Many agricultural programs in China have been wrecked even before they were started simply because of overstaffing. Many of the previous national attempts to develop Chinese agriculture have been thwarted by forcing the people to fit into an organization rather than adjusting the organization to the needs and desires of the people. Such mistakes are all too common and need no specific illustration. The point to be stressed, however, is that all governmental agencies should be created for the people and not vice versa. Here lies the secret of democratic government. To make extension effective in China, it is imperative that those serving in the overhead structures should not lose sight of the dirt farmer who tilled the land, harvested the crops and created a great civilization long before these public agencies came into existence. It is axiomatic that governments may come and governments may go, but the people remain forever. After all it is the people who count. This truth applies to extension work as well as to public administration. Every extension effort should put people first and simplify and consolidate the overhead structure as much as possible.

Planning with a Deep Conviction and a Strong Faith. There is one highly important maxim for the field of agricultural extension; one must give himself or herself to serve the rural people with a deep conviction and a strong faith in agriculture and in the rural people themselves. Extension is designed to build rural people as well as to improve agriculture. According to Dr. C. B. Smith, "Extension is an educational agency that not only helps rural people increase their efficiency and their income, but also helps build rural people themselves into understanding, accomplishing, self-confident, capable men, women, and youth—rural people of vision and leadership. This building of rural people is the ultimate end of Extension." [4]

To serve rural people, one therefore needs to have a deep conviction and a strong faith. It would be fatal to this profession if the extension workers merely regarded it as a cloak put on for political or campaign purposes. On the contrary they should have a deep, abiding faith in the inherent possibilities of agriculture and the intrinsic value of country life. Indeed, the success of any agricultural reconstruction, and Chinese national reconstruction for that matter, will largely depend on an alert and well-informed rural people, jealous of their rural heritage, serious about their own rights. Many Chinese leaders propose to build up their nation through the channel of an Extension Service whose ultimate objective is to build men.

CONCLUSION. Today Chinese agriculture, like Chinese society, is undergoing a series of changes and adjustments which are being demanded by the great majority of the Chinese farmers. To bring about these changes, China needs to establish a solid agricultural foundation beginning from the bottom. It would serve her national interest well if the Chinese Government could help her people to build right from their rural neighborhoods and communities. No nation can really flourish if her people are chained to poverty, ignorance, and disease and are constantly under the grip of economic exploitation and political abuses. There is no short cut to national regeneration. However, one can recommend with assurance a cooperative agricultural extension system which is essentially an educational process of teaching, suggestion, and persuasion calculated to change the working and living patterns of Chinese farmers and to bring about a better economic condition which promises the people a higher standard of living.

How can one make extension effective in China? To be sure there are no iron-clad rules. It needs to be reiterated that, taking the United States as an example, extension is based on cooperation, education, and local leadership. As Director M. L. Wilson once put it:

Extension work grew out of the needs of rural people. Its record for 30 years has shown it to be an instrument for good which rests primarily in the hands of rural people. Extension work is not an instrument of pressure

[4] C. B. Smith, *What Extension Is* (Dept. of Agriculture, Extension Service, Washington, D.C., 1944), pp. 3–4.

groups or an agency for which any political party can claim special interests. Its success goes back to the principle of cooperation, which was basic in its founding.[5]

Extension has never been enforcement, as far as the American system is concerned. It is democratic in and through all its operation.

In introducing this system into China, let extension workers beware trying to do everything too quickly. Play safe with the farmers. Agricultural reformers gain little by stirring up discontent among the peasants unless they are sure of their ground and are able to do something to meet the felt needs of the people. It also does more harm than good to start with grandiose schemes or proposals and force them down the throats of the people. And lastly, let the Chinese extension workers talk little about calories, much less about vitamins, but use the common man's "lingo" to demonstrate the plain value of a square meal with a bowlful or two of the whole grain rice. The bowl with rice is the Chinese consolation! To give the people enough to eat should be the main immediate concern of the Extension Service.

Finally, let this simple truth ever stay with the Chinese leaders; namely, that no one can live exclusively unto himself, as far as modern agriculture is concerned. China cannot attain these objectives entirely by her own efforts. Help from and cooperation with other countries are indispensable. To follow this truth is to reemphasize the concerted viewpoint of those who participated in the Hot Springs Conference, where leaders from different countries pronounced that "the goal of freedom from want of food, suitable and adequate for all peoples can be achieved . . . that the primary responsibility for doing so lies with each nation . . . but each nation can fully achieve its goal only if all work together."

[5] M. L. Wilson, "The Next Thirty Years," *Extension Service Review* (Dept. of Agriculture), XV, No. 7 (July, 1944), 98.

Chapter 6 · EXTENSION EXPERIENCE IN INDIA · By D. Spencer Hatch

FROM NORTH TO SOUTH and east to west the people of India, numbering almost 400 millions,[1] constitute a great variety in physical and racial make up, in religious and social customs, and in castes and creeds. Eighty-six per cent of the people are rural, living for the most part in 655,892 villages, which average less than 500 inhabitants each. The people go out from the villages to work the surrounding fields. Whereas in the old days this congested living afforded a measure of safety, for the villages were often walled, this protection is no longer necessary except in some parts of the north. Such congestion, of course, makes for less sanitary and healthful living. In some areas there is fortunately developing a trend toward living on the land, which means that some families are moving out of the villages and establishing farmsteads of their own.

It is impossible in brief space to describe or generalize about the great variety of land tenure systems of India. One can say, however, that in most parts of the country the land has become increasingly fragmented because the father divides his land among his sons. This means that in spite of the extremely small holding a family may not have all of its small pieces of land together. This fragmentation, or parceling, of land is one of the most difficult problems of rural India, but, in spite of its baffling nature some of the Provincial governments have started to work on the reapportioning of farm holdings.

It is estimated that there are 2.2 acres of land per family, an area which requires the most intensive and expert cultivation to feed the family, but which is generally worked according to the poorest methods. One seldom sees a field that is producing what it should. Furthermore, there is a vast amount of land not cultivated now but which could be used if irrigated.

The village people, as a rule, go on cultivating the exhausted soil by

[1] According to the census of 1941, India has a population of 388,997,955—an increase of fifty million in ten years. 49,500,000 live in urban areas and 339,500,000 in rural areas, or a ratio of seven to one. In the provinces of British India the population is 64.5 per cent Hindu, 27 per cent Muslim, 1 per cent Christian. For all India, including the Indian States, there are 66 per cent Hindu, 34 per cent Muslim. In the two states where the author has had his headquarters there are more Christians: in Travancore, 32 per cent of the population, and in Cochin, 29 per cent.

the same inefficient and laborious methods used for centuries past. This is because the Indian farmers today know the art of farming only as their forefathers did; they have not advanced in knowledge as have contemporary farmers in most other countries, and they still use primitive implements because better ones are not available to them. Long dry seasons enforce idleness, which in turn dissipates the working hours of an otherwise potentially great manpower; tradition imposes economic prejudices and the fragmentation of the land; poverty licenses the ubiquitous moneylender; fatalism clothes life in a hopeless apathy. Literacy was 7 per cent in 1931 and 15 per cent in 1941, although the rate for the village population is said to be not more than 2 per cent. It is no wonder, then, that India is in need of and ready to receive any offering of good will and service.

CHARACTERISTICS OF THE PEOPLE. In spite of the great diversity of living conditions, rural life in India, as in other Eastern countries, depends much on the land, family, religion, and caste or community. From these come certain common traits which can be enumerated as characteristic of Indian rural people.

The fare of the home, whether meager or more adequate, is always shared freely with the guest.

Almsgiving is practiced to such an extent that it encourages begging. India probably has more than its share of beggars, uncounted thousands who spend their lives in the expectation that people will give to them either from generosity or because they believe that thereby they store up merit for themselves in heaven.

The patriarch, who is the father or eldest male member of the family, is greatly revered. Other members do not eat before he comes nor do they sit until he sits. His experience and wisdom are truly appreciated and relied upon. In most communities the boy and girl have little to say about selecting a mate for the parents look for the most beneficial and profitable match, and the parent of the boy can dictate to the parent of the girl, from whom he collects a dowry. With the coming of education this practice is often changed.

The rural people are beset by fears of countless kinds: fears of terrible gods and goddesses, of diseases and plagues and snakes, failing crops, and hunger. In addition, they fear creditors, the scheming person, and the robber.

Willingness to put off until tomorrow what could be conveniently done today is general. The words for "tomorrow" are the ones most frequently heard in the two hundred languages spoken in India. The attitude toward life is leisurely.

Hopelessness is an outstanding characteristic and is the inevitable result of the frustration continued from generation to generation. The Indian rural people would rather follow than think, because they are tired of fruitless thinking. They have also developed as a result of their hard lot a superstitious complex in their dealings with other individuals, but underneath they are a frank, trusting, communicative people.

Their whole culture is precious to them, especially because it is one of the few things they have succeeded in keeping intact in their struggle for survival and freedom against a tragic background of poverty. Most villagers are capable of change and will respond to the teachers whom they trust, even though their past sad experiences make them conservative and hopeless about the future. They are eager to learn how to help themselves and they resent paternalism. They borrow money at exorbitant rates of interest, for they have no other choice. They welcome the establishment of small local industries that will provide income to supplement the poor living off the impoverished land. They see that giving their best to farming often results in enriching others and leaves them just where they were. Poverty and fear of extinction have given the Indian villagers a kind of doctrine of live or die together, a form of cooperation weakened only by religious and caste differences and sometimes by local feuds. Many a prophet in India has come from an Indian village.

THE NEED FOR RURAL RECONSTRUCTION AND EXTENSION. The foregoing paragraphs are not in any way an adequate description of the backward condition of the rural people of India. There is a wealth of good books on this and related subjects, which anyone interested can read and study. No one will doubt the crying need of extension teaching and demonstration. And these are peoples who have to be shown. The demonstration method succeeds where mere talk is just more talk. Talk is the cheapest and most universal indoor and outdoor sport in India.

Although the villages are pitifully behind the times, deteriorated and disorganized, they play a most important role in the economy of India. They support the country to a great extent and yet are themselves practically starved. Many conquests, both domestic and foreign, have through the centuries contributed to this impoverishment. Heavy outside demands have greatly weakened the village economy. Most of the advantages of the government and of private agencies have been enjoyed in the towns and the few cities. The villages have been left helpless. Progress, as measured in the development of industries and education, has stayed mostly in the towns.

Overpopulation is often considered a leading problem in India. It would not be such a worry if the land were made to produce the food which it could provide. The theory that population increases with food supply need not apply to India where education quickly cuts the birth rate and where other checks to population growth operate. One serious problem does deserve considerable attention: the plight of uprooted peasants who migrate to the city because of village poverty and the lack of small village industries. These migrants present a terrible picture of degradation as they crowd together in the worst imaginable urban slums.

The villages suffer too from the draining off of their best potential leaders by educational institutions. After their training, these one-time peasant boys and girls do not want to return to their villages. In fact, more than 90 per cent of the agricultural graduates of India take up civil service jobs in which they can earn more money and thus reject the occupation of farming which has not yet risen into its rightful dignity.

Most villages use cow manure for fuel when it is badly needed for fertilizer. But no other fuel, such as coal, oil, or wood, is available. Lack of roads hinders shipment of produce. Lack of refrigeration, even on the railroads, lets perishable foods spoil. Although villages are the centers of food production, the cities buy up, store, and export the grain, with the result that the villagers have to run to them for food during famines. At such times, peasants, in their weakened condition, frequently die in the cities where they have gone in search of food.

India has 250,000,000 head of cattle, almost half of the world's total. Yet there are only four cups of milk a month per person. The underfed, milkless herds are an economic drain on the country.

There are other outstanding needs of the Indian farmers. Several of these were drawn up at a recent conference by a group of twenty persons, all of whom had spent years working in different sections of India.[2]

Increase of land area under cultivation through irrigation.

More adequate means of communication, as by improved roads, railroads, and radio.

Reforestation by encouraging each village to plant as many trees as possible, year by year, and by protecting young trees from grazing cattle.

Readjustment of land tenure systems in favor of the farmer.

Discouraging absentee landlordism, as by facilitating land ownership by the farmer.

Improvement of soil under cultivation by providing fuel to replace waste materials now used as fuel but needed for fertilizer, and through provision of commercial fertilizer.

Improvement of crops, both in quantity and quality, by the sale of young trees from controlled nurseries and the sale of improved seeds for vegetables as well as for staple crops.

Improvement of stock through loan or sale of bulls of selected breed, sale of eggs of better breeds of fowls and grading up with purebred cocks.

More efficient use of labor and time through preparation of work schedules.

Encouragement of small industries which can be conducted profitably in a village community.

More profitable marketing, as through cooperative selling societies and by encouraging traditional weekly markets.

Reduction of indebtedness through expansion of cooperative credit societies, by increasing cash earnings, and by urging governments to make the charging of exorbitant interest rates a criminal offense.

WHAT THE GOVERNMENTS OF INDIA ARE DOING. Provincial governments and those of the Native States have agricultural departments and make use of various methods, more or less successful, to spread the results of their investigations. On the whole, the governmental activities have been much more adequate in the field of agricultural research and experimentation than in the extension of the valuable findings to the really poor rural peoples. This observation has been voiced by many observers, including Sir John Russell, in assessing the work of the Imperial Council of Agricultural Research. The

[2] Conference to Outline the Contribution of Extension Methods and Techniques to the Rehabilitation of War-torn Countries, September 19-22, 1944, Washington, D.C., sponsored by the U.S. Department of Agriculture.

projects of an extension nature carried on by most Provincial and State governments include:

Seed stores, usually in strategically located market towns. These are used as outlets for improved strains of seed and for the sale of improved implements.

Demonstration farms, intended to encourage improved agricultural practices through example, but actually most useful in conducting varietal tests on various crops.

Veterinary service, contributing chiefly through promoting castration of scrub bulls and through combating epidemics of animal diseases.

Sponsoring cooperative societies, through government organizers who set up societies and arrange for periodic auditing of accounts.

Encouraging village industries, chiefly of a handicraft nature.

Public health services, chiefly smallpox vaccination and distribution of quinine.

That government officials realize the utter inadequacy of these measures is demonstrated by their concern with integrating all governmental services to rural people, by their intention of expanding greatly the provision for rural extension in governmental budgets in postwar years, and by such official pronouncements as the Viceroy's message to the Government's Postwar Policy Committee on Agriculture, Forestry, and Fisheries (June 26, 1944).

MAKING EXTENSION WORK EFFECTIVE. Because of this quickening of interest in rural extension work on the part of both public and private agencies in India it is timely to point out what facts contribute most to the success of such endeavors. Many programs designed to help the villager will be launched with enthusiasm, but many of these will fail because they do not take into account some of the "pillars of policy" which should be an integral part of the programs. The author, because of his experience of over twenty years in India, and particularly with the Rural Demonstration Center in Martandam, Travancore, is convinced that many difficulties can be avoided if certain fundamental principles or guideposts are realized in advanced. In the discussion and illustration of these principles most of the examples will be taken from the files of the Martandam Center, which was started in 1921 under the sponsorship of the Indian National Council of the Y.M.C.A.

Work carried on by others in different parts of India offers abundant proof, in stories of both success and failure, of the validity of these principles. As they become more generally recognized and more widely applied, they will contribute to the rural uplift so urgently needed in India.

Selection of the Site. Extension work must begin somewhere whether it is carried on from a definite center or as part of a broad educational program by governmental officials. The choice of the place is of strategic importance. One thing has proved true through the years, both in India and in the "poor man's cooperative" of Denmark: that is the willingness of the rich to copy the poor and the inability of the poor to copy the rich. If one helps the poorest on the poorest land to help themselves then those better off can be persuaded to copy whatever seems to work for those beneath them. Far too often, those with more education and larger economic resources have appropriated the benefits while the poorest have been left out. But by careful attention and planning the poorest can be helped; alone they can do nothing. Martandam was chosen for this very reason. The land was much poorer than other parts of Travancore State; there was less rainfall. It was far away from the cities but yet a motor road, with buses, made transportation of products possible. The villages in the surrounding area contained representatives of the many kinds of people in Southern India. A clinching argument for the choice was the big rural high school of the district, a point which later proved of extreme importance in the success of the Center.

Also of importance in selecting a site is the attitude of the people toward the proposed work. Experience has shown that, where the need is so great, it is best to work only in those villages where people really want help and invite the extension workers to come. When work succeeds in one village it spreads like contagion to other villages.

Understanding the People. It is a great mistake to work anywhere in any country with just one kind of people or to let one group feel that they are preferred above all others. This means that those responsible for extension work should become well-acquainted with the social distinctions among the people whom they are trying to serve. Often, some nonstereotyped survey which would not worry the country folk is sufficient for the purpose. In the case of Martandam, within a three-mile radius there were slightly over 30,000 Hindus, a little less than 10,000 Christians, 600 Mohammedans, and 1,000 of other faiths. These were again divided into small groups, no one of which liked the other too well. The Roman Catholics and the London Mission Christians had a minimum of dealings with each other. The Hindus and Mohammedans were not overfriendly. The Hindus themselves were divided into eleven main castes. Teaching these people how to work together in any program of village betterment has proved no mean task, albeit a necessary one. Basic to such cooperative activity, however, is the knowledge of what groupings exist and how these groupings already get along with each other.

But understanding the people also means knowing about their ways of life—their religious beliefs and practices, their family system, their attitudes toward economic matters, their standards of right and wrong —and a number of other things which, if violated, would jeopardize any extension program or which, if utilized, would speed the program along.

The degree to which beliefs and other culture traits can interfere with extension programs is illustrated by the poultry-keeping project at Martandam. In order to improve the flocks and to make poultry keeping more popular, a religious taboo had to be faced frankly. Many Hindus simply would not eat eggs because in so doing they killed the life germ in the egg. To overcome this cultural obstacle, "vegetarian eggs" had to be put on the market after the people were shown the impossible—that unfertilized eggs had absolutely no sign or possibility of life any more than did milk from sacred cows, and that they could be eaten with the greatest safety, assurance, and profit.

Once this hurdle had been taken the project got in full swing. Eggs were given to village leaders who were ready to spend enough time at the Center to learn how to care for the eggs and the chickens when hatched. The pure-bred fowls which resulted were propaganda enough. Then a Cock Circuit was organized. Any village family who would get rid of their common cocks could have a pure-bred cock for two months, during which time they had to set whatever eggs were laid. Another development was the establishment of the simplest cooperative egg marketing which, in a sense, rounded out the poultry project.

Formulating a Well-adapted, Comprehensive Program. A program tailored to fit the needs of particular communities often sells itself, especially if local people have had some part in the tailoring process. It makes possible the condition which might be described as the "suction" rather than the "squirt gun" situation. This is a necessary condition for good extension work and means that the people are asking the workers to come out to help them. It is unlike the approach of some experiment stations which have developed something which they think good for the farmers and which they try to "squirt" at them. People are naturally shy of squirt guns. At times, as in the case of the author's present experience in Mexico, the "suction" situation occurs almost before it can be managed, and the problem resolves itself into providing enough extension workers with sufficient expert knowledge to take care of the countless requests. The popularity of a program therefore, depends in great measure upon local recognition of the need for it. It must seem vital to the people; sometimes great patience must be shown in teaching them the importance of the program and its application to their daily lives.

Not only must a program be "tailored" so that the people feel that it belongs to them, it must also be comprehensive. This is perhaps the foremost pillar of policy in rural reconstruction work. Someone has written that "this discovery of the comprehensive program bids fair to be one of the most valuable contributions to rural science in this generation." Well, maybe it is a discovery. Anyway, experience in India taught that in trying to help those who were pitifully poor there was little good in meeting the farmer's needs along just one line. For instance, not much was accomplished by a five-year plan or program which dealt only with health. Hospitals around Martandam discovered that all the time that medicine was being given out, a good part of the people were so hungry that they could not possibly be healthy. New methods of agriculture, sanitation, or hygiene were ineffectual unless one or two cottage industries filled the spare time of the idle people enabling them to increase their bartering power or purchasing margin.

The purpose of the work of the Martandam Center was finally defined as follows: *The purpose of rural reconstruction under our Association is to bring about a complete upward development towards a more abundant life for rural people, spiritually, mentally, physically, socially, and economically.*

Not only did the program make its influence felt in all phases of life, but it came to embrace all age groups and all castes and creeds. For example, it did little good to teach boys and girls to eat unpolished rice, sleep with fresh air, or build better poultry houses, as long as grandmother ruled the house with an iron hand and knew better than young "upstarts" what ought to be done. So the comprehensive program had to gather in grandparents and parents. Any kind of cottage industry meant that a young boy had to have the help of mother or sister if he were to succeed.

A further characteristic of a sound program is its long-time point of view. One cannot accomplish anything permanent by placing workers in a needy rural area for a short time only. It is the American way to want to go more than twice as far on less than half as much, but the Indian villager is conservative and slow; he has to be shown. Furthermore, he must have his counselor stand by through his trials year in and year out. A rural reconstruction center is a life work, which always builds, grows, and develops. Work that is started should be of a permanent nature; only thus can it have a far-reaching influence. Trained leaders should not be moved so rapidly from one area to another that they become merely "rolling stones." In the great needs that agencies will face at the close of the war—and are even at this moment facing in liberated areas—the temptation will be to look upon leaders as "emergency kits" rather than master builders. Much emergency work will be necessary, but one should always keep in mind the permanent element.

Another feature of program planning has been proved essential through years of experience. A program should build upon something which the people already have, with which they are familiar. It should not seek to introduce a product or a method which is entirely new and so foreign to the villager's experience that he has nothing to which it can be related. For instance, hand weaving introduced into an area where there is absolutely no weaving or no tradition for it has generally failed. This is true even though the people may sadly need such a good home industry for their spare time. On the other hand, where some weaving already existed great strides could be made in its development by the improvement of methods, setting higher standards of quality, introducing greater variety of patterns, and by assisting in the marketing of the product.

Beekeeping affords another illustration of successful building upon what the people already had. Honey had been used in India for centuries as medicine rather than as a food. The Ayurvedic physicians, following their centuries-old formulae, bought up all the honey they could find and sold it to their patients who dripped it drop by drop from narrow-necked bottles. Beekeepers were few and even these hung old cracked cooking pots in the trees and hoped that colonies of bees would come to live in them. Some did, but the life of bees in India is precarious: snakes, ants, lizards, and other marauders drive them from home to home.

Workers from the Center started beekeeping on modern lines in hives specially adapted to the smaller size of the Indian bees. These bees were transferred into the new hives from old pots, hollow trees, caves, roofs and wherever they could be located. By the use of a small extractor, clean, delicious honey was placed on sale at the Center. The excellent price that this honey brought was the chief talking point with the great number of potential beekeepers who sprang up on all sides. A number of men and boys helped at the Center long enough to learn about the care of bees, the transferral into a new hive, the extracting and bottling of honey. The government, seeing the great possibilities, gave timber from the forests for hundreds of hives which were made at the Center and sold to the beekeepers at cost price.

This beekeeping spread miraculously. A testimony to the value of simple demonstration was given by three brothers who came to one of the exhibitions with a very large exhibit of honey as well as their homemade equipment. They were from a distance and were not known by anyone at the Center. When asked how they learned the modern way of beekeeping they answered: "When you were having classes in beekeeping at the Center we just stood in the crowd day after day and listened and watched until we learned very well." These brothers were earning a sure income for their families from the considerable business they had developed. As an outgrowth of this widespread interest in beekeeping a cooperative society has developed. This association employs a secretary who manages the extraction and sale of the honey which, under the cooperative's label of quality, is a product much in demand. Honey is still a medicine for the majority in India but for hundreds around Martandam it has become a relished food.

Making Use of the Right Kind of Personnel. A good rule in extension work is "Never do anything yourself that you can get someone to do for you." This calls for the development of local or volunteer leadership. In the work in Martandam the limited budget permitted the employment of only two or three helpers, so great reliance had to be placed on "honorary" workers. A recent count numbered these at 1,310. Rural reconstruction must be a self-help movement, one in which the villagers are helped to help themselves. Paid workers have an important role to play, but in the final analysis the project must be done by the villagers. It must be their choice, something they want to do. Only then will their interest carry them through to success by exerting the pressure of public opinion upon those who are recalcitrant or slow to adopt the new practice accepted by the majority.

The lowly borehole latrine is a case in point. Crowded villages, with houses almost touching each other and with absolutely no sanitary systems, meant a host of barefooted people walking in human filth. Some villages tested one hundred per cent hookworm. The borehole latrine seemed the most effective means of combating this menace and of promoting sanitation generally among the people. When workers from the Center first mentioned this idea to various village leaders they replied quite frankly, "This sounds ridiculous. We don't need anything like that. We have open space around our homes." A few leaders, however, put down latrines. Only a few months later these same disbelieving villagers were making more eloquent and impassioned speeches than any paid worker as to the need and value of the borehole latrines. Then came the time when the majority of the families in a village had latrines. Those who had not adopted them began to receive public censure and ridicule as "dirty people." Gradually these people complied and the village became "completely sanitated." But more than that. The borehole latrine soon became hailed as a profitable cottage industry because people found out that by using the double-hole system the contents of the first hole, completely pulverized, decomposed, and deodorized by the time the second was full, could be sold for fertilizer at the princely sum of seven rupees.

In spite of the low literacy rate for all India, Indians do have a real thirst, indeed a great thirst, for learning. In Travancore State the Education Department enrolls thousands of rural teachers, who comprise a large percentage of the honorary workers of the Center. Farmers, lawyers, bullock cart drivers, women, even boys and girls, all contributed of their time and energy. Any professor who had something to contribute was invited from the colleges located in the city, twenty-five miles away. In such an all-round program, help of many kinds was needed.

Although most of the governmental authorities and specialists lived in the city they contributed their expert knowledge to the program. They seemed to appreciate the way in which a nonofficial agency could bridge the gulf between them and the people, with the result that the good things they had to offer could actually reach the poor who needed them. The governments of India, both Provincial and State, have the same difficulties that one finds in other rural areas— the rural people look with suspicion upon governmental programs. They say, "If the government makes us more progressive they will tax us more," or "If the government wants to do anything, let it come and do it." Without the cooperation of local people government programs accomplish little in the long run.

In addition to local leaders and available specialists, any extension program does call for trained personnel to work full time with the rural people. Selecting the man and woman for such work is of tremendous importance. For one thing, a rural background is definitely needed. City men are sometimes sent out to reconstruction centers or placed in extension work and it takes years of service to transform them into useful rural workers. They frequently lack any sympathetic understanding of rural people, any love for the soil, any liking for the simplicity of rural living. Leaders who have been brought up on a farm thus have a real advantage. Another difficulty is that formal education, in spite of its great aid, sometimes trains rural students "away from" the village. One outstanding student sent by the Government of Ceylon for training at Martandam is a good case in point. The first morning at 5:30 he reported for practical work dressed as a government official dresses—silk shirt, silk suit, expensive tie, polished shoes. He began with the others to learn by doing, digging in gardens, making compost, caring for the animals. The second morning he folded away his officer's clothing for the duration of the training school. Soon he was out in shorts and sandals, playing games with village children. He said quite frankly that since he returned to Ceylon from study in England he found that he could not talk with village people among whom he was sent to work. There was a barrier; he was so different from them. And he ended by saying, "Now I know why. I *was* different." He finished the Martandam course with honors and the reports from his subsequent work for the government in rural Ceylon are records of outstanding success.

Methods Must Remain Simple. Simplicity must be the keynote in all efforts with underprivileged rural people. The worker needs to demonstrate simple habits in his own life and to be willing to teach and help people to practice only the simplest and least expensive methods. Rural reconstruction workers need to be rich in the things they can do without. For instance, the buildings at Martandam can be easily and cheaply copied. They are structures of mud walls, bamboo, and coconut thatch, made to house efficiently and sufficiently the demonstration animals, the seed bulls, the milking goats, and poultry. The library cost only twelve dollars, yet its one locked room serves as a repository for the books; its roofed-hut, open reading room is the meeting place for a great number of village committees; and its uncovered verandah overlooking a volley-ball court is a stage for dramas presented before delighted audiences who squat on the court. This verandah is also a platform for public speakers and for the awarding of prizes.

Less than twelve dollars was needed for the substantial lean-to protecting the eight looms of the weaving school and the blackboards of the night school. An ingenious hut, headquarters for the Boys Work Department, cost nothing but the labor of the boys who gathered the bamboo framework from near-by jungles and who planted antiganon vine for the walls. These vines, incidentally, offer pasturage to the bees whose hives range in stately splendor behind the hut. Paddle tennis courts provide space for nonequipment games. The requirement for a nation-wide rural reconstruction program is many trained leaders. Martandam, in its Practical Training School of Rural Reconstruction, has trained over 1,000 picked men and women to go back to their home areas in different parts of India, Burma, Ceylon, China, and Egypt. Yet the main classroom building of this educational institution, which some have called a "university," is thatched with coconut leaves and cost only the equivalent of seven American dollars.

One of the most effective methods among rural people is that commonly known as the demonstration method. People can see how a thing should be done. Although demonstrations were staged occasionally at Martandam the whole place was in effect a large demonstration center. The 1,500 schoolboys and girls next door could ask as many questions as children anywhere in the world. They asked for eggs and seeds and beehives. The swarms of people passing the gate on market day would often rest their head loads on the fence posts or come in to squat under the shade of the trees to rest and watch the new "goings on." When they saw something within the reach of their understanding and financial ability to copy, they had it. Sometimes they requested help, sometimes they went silently away to work out their ideas in their own way.

A question frequently asked by those interested in the work of the Center was "How do you get people so divided to work together?" Division is a serious problem upon which hinges much of the poverty, weakness, and lack of accomplishment. In every village there are those too poor and too few in number to do things alone. The divisions among the groups were so strong that it had become almost traditional for other groups to work against anything one group tried. If a Christian proposed a road for a roadless village, it was dubbed as a Christian movement and others could not help. If a Hindu proposed a well for the village, that was a Hindu movement and, no matter how thirsty he might be, no one else would help. Mr. Gandhi and other great and trusted leaders like him had spent a good part of their lives on this problem, and were discouraged by their lack of success. They often tried mass schemes. At Martandam the approach was just the opposite —with the individual. As soon as certain enlightened leaders in each group understood the nature of the program, they saw that only through cooperation among the groups could the village rise and be strong. Many of the leaders, in fact, deplored the tendency of groups to oppose each other and the part they had to play in this opposition, but they were the victims of the deeply-rooted system and knew no way out. Being nonsectarian, the Y.M.C.A. Rural Demonstration Center offered a common platform to which rallied many leaders. Without losing face in any way, the more adventuresome were ready and eager to join their people with others for the common good.

The committee system worked well. If a whole village were to embark upon a project for the benefit of all, representatives of each caste and creed had to decide upon ways and means of carrying the project through. As they gathered around the committee table to discuss their problems, these leaders put aside the prohibitions of their castes and the distinctions of their creeds, for they were intent only upon accomplishing the desired and needed ends. It has always been a thrill to plan and work with a committee numbering, say, a Brahmin, a middle class Hindu, a backward class (not long since an outcaste), a Protestant and a Roman Catholic Christian, a Mohammedan, and a Buddhist. When these are willing to unite, what avenues can remain closed? Such committees enlisted helpers from all communities. The number of roads opening up roadless villages (some of them with laborious revetments and excavations) and the number of wells in waterless villages mount yearly to a creditable total. Roads mean the transfer of heavy loads from heads to bullock carts; when a well comes to a village many steps are saved for those steadying brass water pots on their hips. But more important than these physical and material achievements are the results in cooperation. Cooperatives formerly organized solely for one religious group—Christians, Nayars, or Brahamins—now are becoming mixed, as the government policy toward licensing segregated cooperatives undergoes change. In a sense, this is a spiritual gain.

The Foundation for Rural Reconstruction Is Spiritual. In countries on both sides of the world rural people are particularly spiritually minded. Their lives are bound up with the simple natural processes. They are free to the open world and are not warped by the man-made complications of urban living. It matters not the slightest what religion a rural dweller professes—whether Hindu, Christian or Mohammedan. He feels that all helpers in extension must have a spiritual basis and tries to find such a foundation in his own religion. Or he seeks for it in other religions. He realizes that officers who are careless in religion cannot lead spiritually minded people.

THE IMPORTANCE OF LITTLE SCHEMES. Because India is so big and its rural population so tremendous it is but natural that the governmental schemes for improvement should prove likewise gigantic. There are grandiose postwar schemes for water catchments, irrigation, research institutions, roads, and railroads. Most of these lie within the field of government and must be done by the government if done at all. But they will not solve India's problems. The central feature and problem of Indian rural economy is still the small scale of both the operations and the resources of the Indian farmer. A professor friend has written, "India is par excellence a country of small peasants. Universally the problems of Indian agriculture are those of the small peasant and must be visualized from his point of view and solved in the light of his needs and circumstances."

This is why a sound extension program must be within the range of the meager resources and fragmented holdings of the average farmer. But such little schemes become big when multiplied by the efforts of thousands of small groups working on them. The big scheme which will work best for India is this multiplication of small efforts. Such efforts move forward as basic extension guideposts are followed and as local leaders are trained for positions of responsibility. Only in this way will they become truly "popular," or of the people.

This is primarily the story of extension experiences in India. Here and there a note of Mexico has crept in because the author is working on rural reconstruction there. It is interesting to see how work on opposite sides of the world takes on so similar a cast. When life becomes a struggle for existence it speaks a universal language. The ways and means to meet the needs are universal also.

Chapter 7 · EXTENSION WORK AMONG THE ARAB FELLAHIN · By Afif I. Tannous

THE REGION AND ITS POPULATION. There are six political units in the region under consideration—Egypt, Palestine, Transjordan, Lebanon, Syria, and Iraq. Each of these has its own legislature, government, customs barriers, currency, foreign policy, and similar prerogatives of statehood. In reality, however, such points of differentiation and separation are only skin-deep. Underlying them are the basic bonds of a cultural background that is predominantly Arab—the religion of Islam, the Arabic language, and an agricultural way of life. Consequently, the problems, principles, and techniques of extension encountered in any one of these countries are similar to those encountered in the rest of the region.

The total area is rather extensive, amounting to about 685,000 square miles. Most of this, however, consists of arid deserts and semiarid plateaus over which the nomadic Bedouins graze their flocks. The area that is normally under cultivation does not exceed 45,000 square miles. Within this limited space lives most of the total population of 28 million, concentrated mainly in the Nile Valley of Egypt, the Tigris-Euphrates Valley of Iraq, the littorals of Palestine, Lebanon, and Syria, and the oasis of Damascus. Egypt alone has a population of 17 million. Density of population per unit of cultivated area is about 550 per square mile, reaching a maximum of 1,000 and 1,300 in Lebanon and Egypt, respectively, and a minimum of 70 in Transjordan and 80 in Iraq.

The main population aspect which is of significance to our topic of discussion is the rural-urban composition. First there are the tribes, consisting of nomadic or semisettled Bedouins. For countless centuries these migratory groups have roamed over the plains and plateaus of the region, following the seasons with their flocks of sheep and goats. Their fundamental role in the total way of life in the area is indicated by three main contributions. From their overflow of population, village settlements have been increased and urban centers replenished; by the surplus of their livestock products, agricultural production has been supplemented; and under the influence of their tribal organization many of the traditions, folkways, and mores of the local culture have developed. At present they number about four to five million people in various stages of transition from pure nomadism to settled agriculture.

Not more than seven million people are considered urban. At the same time many of the centers classified as urban come directly under the influence of the rural way of life to one degree or another. By far the greater portion of the population, about sixteen million, are village dwellers who depend almost completely upon agriculture for a living. This is the society with whom agricultural extension is mainly concerned, and whose way of life will now be presented briefly.

THE VILLAGE COMMUNITY: *Physical Structure.* Unlike the pattern prevailing in the United States, there are normally no isolated farmsteads in the Middle East. All cultivators of the soil, usually called *fellahin* (singular, *fellah*), live in villages and go out to work in the surrounding fields as the occasion demands. Almost invariably the villages are of the nucleated type, with houses built close to one another, around a mosque or a church as a center. In each conglomeration there may be sections (each of which is called *hara* or *heiy*) that are identified with various kinship groups. Within the settlement, usually close by the place of worship, there is an open space (*saha*) which serves for trading and social gatherings. In view of the semi-arid climate of the area, the source of water supply for drinking or irrigation is an essential factor in the physical structure of the community. This may be a stream, a spring, or a well. Garden plots may or may not be available within the village proper, whereas all around it lies the agricultural land, consisting normally of open fields, orchards, and grazing grounds.

The Pillars of Its Culture. There is in the first place the land. It is the firm foundation on which village life is anchored. Its significance goes beyond its economic value. It figures strongly in many traditions and is the object of strong emotional attachment. Two main aspects of this pillar are land tenure and agricultural activity. Four principal types of land may be distinguished. *Miri* is probably the most predominant category. This is state domain, rented out to various tenants for differing periods of time or leased indefinitely to a farmer and his progeny against the payment of a regular tax. Other conditions are attached to this form of ownership which tend to hamper agricultural progress. *Mulk* is land that is owned in fee simple, entailing practically absolute rights with respect to disposal and manner of cultivation. In addition to *Miri* and *Mulk*, almost every village has some property that is dedicated in perpetuity for religious or charitable purposes. This is called *waqf*. Finally, in many villages a form of communal ownership prevails, which is called *Masha'*. Members of the community own certain shares in the territory, but no specific plots. Allotment of these for cultivation by each family takes place at regular intervals.

Special mention should be made of one aspect of land tenure that has direct bearing upon the possibility of extension development in the Middle East. This is the fact that the great majority of the *fellahin*, possibly about 70 per cent, cultivate the land as tenants or share croppers of one sort or another. The real owners are the tribal Sheikhs, the feudal lords, and the absentee Effendis. In many cases these own the land, the village proper, and, virtually, the *fellahin* who live in it.

It can be truly said that agriculture among the people of this region is a way of life. It is stable and deep-rooted, and its techniques are the product of countless generations. To the *fellah* there is no segregation of his land and agriculture from his religion, social activity, family life, and community organization. Another feature is the prevalence of ancient methods of cultivation. The heavy hand of the past can be clearly seen today in the sickle, the threshing board, the team of oxen, the wooden plow, and the chicken coop. Each of these has a long history behind it, is related to other aspects of the culture, has an emotional value, and in many cases is beautifully adapted to local conditions.

Family constitutes the second main axis of life in the Arab village. Within its organization three units or stages can be recognized, the biological unit, the joint family, and the kinship group. The first of these consists of the married couple and their children, and the scope of its function and significance is relatively limited. On the other hand, the joint family (consisting of the paternal grandparents, their unmarried daughters, their unmarried and married sons, together with the wives and children of these) plays a more important role in the life of the individual and the community. Its influence is manifested clearly in such matters as marriage, funerals, conflicts and feuds, agricultural activity, land ownership, religious affiliation, and mutual aid. In most cases the individual is also conscious of belonging to a more inclusive unit, the kinship group, consisting of all those who claim descent from the same paternal ancestor. Normally, a community is made up of a few such blood groups, each of which includes several joint families. In a word, no matter from what angle one approaches the analysis of the Arab village, one cannot fail to observe the fact that blood ties cement its life and describe the behavior of its members to a high degree.

As the *fellah* is born into a traditional agricultural activity and into the bosom of a family group, he is also born a member of the village religious institution, Moslem or Christian. His membership is taken for granted, is never questioned, and is also expected to continue as long as he lives. The idea of conversion is alien to his mentality. His ancestors lived and died along the path of a particular faith, and he cannot see any reason why he should not do so. On the other hand, he has a very important reason, of which he may or may not be conscious, for being averse to making a change in this respect. Religion to him is a way of life, as much so as are his agriculture and family, and discarding one way of life for another is too much of an upheaval. He may not be well versed in the dogmas of his religion, but he readily subscribes to the many rules of behavior it has inspired in social, legal, and economic relationships. This is especially true of Islam, the faith of the great majority (about 85 per cent) in the Middle East. From its early beginnings it took the form of a message which comprised the totality of life and became a dominant factor in Arab culture.

Another significant feature of the *fellah's* church or mosque is that it has little to do with outside religious authorities. In fact, Islam has never developed much of an ecclesiastical hierarchy. The village mosque or church is mainly a local institution, whose building has been erected and maintained by the people and whose property has been donated by them. The priest or *imam* is one of them, whom they elect and support; he is expected to serve them all his life. Unlike the situation in the rural community of the United States, the norm in the Middle East is one church or mosque per village.

With one more axis or pillar, the pattern of life for the *fellah* becomes complete. This is the village as such, which goes beyond land, family, and church. It is all of these integrated together, plus the *fellah's* sense of belonging to something that is bigger than his family and sometimes more inclusive than his church. Through its nucleated structure, this type of settlement affords its members a highly developed level of personal and group interaction that could not be possible otherwise. The voice of the community is heard and normally obeyed in situations involving intervillage conflicts and cooperation, mutual aid, settlement of disputes, moral standards, and public utilities. This influence extends also to such personal affairs as marriage, funerals, rotation of crops, and methods of cultivation. In general each village comes to be known by a certain character or characteristics, such as good or bad morals, honesty or dishonesty, laziness or industry, cooperativeness or disorganization, to name but a few.

Upon these four main pillars or axes—land, family, church, and community—the edifice of the *fellah's* culture has been raised. From each one of them, in interaction and integration with the others, has evolved a pattern of life in which certain values are particularly emphasized. An enumeration of the main ones among these will now be attempted.

1. Generosity in general, and especially hospitality. The guest must be fed, sheltered, and honored. No *fellah* would like to be called miserly. Sometimes this trait is carried to the extreme of wasteful lavishness.

2. Daring and personal prowess. This does not primarily mean physical strength, but rather the possession of what is called "good heart," that which knows no fear in the face of danger. To call a person "coward" is one of the strongest insults. Under the influence of this value there is a tendency to defy authority that comes from outside the village. Going to jail for such defiance is in many cases considered a virtue.

3. Reverence for age. Being old carried prestige and respect. Old folks are cherished and their advice is sought. They possess wisdom, the result of long experience. In addition, the old patriarch is the legal owner of the joint family property.

4. Exaltation of leadership. Normally, leadership is not a birth privilege; it is rather achieved by those who measure up to standards. A community usually has several recognized leaders, each excelling in one aspect of life. Of course, the more desirable qualities a leader has the greater is the scope of his prestige; in general these qualities are intelligence, "wisdom," daring, generosity, land ownership, good farming, learning, and religious devotion.

5. The personal touch in human relations. Personal appeal and mediation may end a bloody feud between two parties when legal procedures, including fines and jail, have failed. A reluctant father usually yields to such personal approach by one or more village leaders and agrees to give his daughter in marriage to a suitor. "For my sake," "for the sake of this beard" are statements frequently heard in interpersonal dealings.

6. Leisurely attitude toward life. Rarely does the *fellah* make use of a watch or clock. He times himself by the sun, the moon, and the stars. Being late to an appointment does not constitute a catastrophe. He uses "tomorrow" to mean the following day, the following month, or the following year.

7. Mutual aid. This is clearly manifested in the manner in which members of the joint family cooperate in cultivating the land, arranging a marriage match, helping distant relatives, and so on. The same attitude is shown by the whole community in such matters as constructing a road, building a church or a school, clearing village wells or irrigation ditches, conducting funerals or weddings.

8. Emotionalism. The Arab villager derives much satisfaction, social and organic, from indulging in demonstrative emotionalism. He is not shy about being sentimental. In times of joy, excitement, or sorrow, inner feelings are expressed freely and vociferously.

THE NEED FOR EXTENSION. During the past fifty years intensive contact between the Middle East and the West has been taking place, with Western culture playing the role of the invader. In general this influence has been felt first in urban centers, especially seaports, and from there it has extended to rural areas. In many cases, however, the contact has been made directly in villages. The return of emigrants, the spirit of nationalism, the introduction of mechanical devices of all sorts, political ideologies, educational institutions, and a rising standard of living have been the principal factors involved in this process of social change. Its degree, of course, has varied from locality to locality, but it can be safely stated that not even the nomadic tribe has escaped it completely. The American automobile and radio are now established in the desert.

Under the impact of the present war the influence of the Western way of life upon the local culture has increased tremendously. The horizon of the Arab village is being widened, the needs of its life are multiplying, its folkways and mores are being challenged, and its traditional self-sufficient economy is no longer adequate. An urgent need has arisen for an extension system through which effective adjustment can be achieved.

Specific Problems. The Arab *fellah* suffers from illiteracy. Not more than 10 or 15 per cent of all village people know how to read and write. There is need for many more schools for children and a literacy campaign for adults. The health situation also leaves much to be desired. Sanitation in the village and within the home is very inadequate. Infant mortality averages about 200 per 1,000 live births. Trachoma afflicts the eyes of not less than 80 per cent of the people. Malaria and typhoid are endemic diseases. The average diet is much below what is considered adequate for the human organism. There is too much consumption of cereals and not enough of meat and dairy products. Fruits and vegetables are eaten only in season, as modern canning facilities are practically unknown in village homes. Furthermore, national economy in these agricultural countries is geared to cash crops rather than to food production.

In his agriculture the *fellah* also faces several problems. The most difficult of these is perhaps his hopeless indebtedness and his continued need for credit. He has to resort to professional money lenders or certain city merchants, who by various manipulations manage to charge him 30, 50, and 100 per cent interest. In presenting his produce for sale, he shows his ignorance of modern marketing techniques. He is at the mercy of the broker or the merchant. His knowledge of how to combat plant diseases and pests is practically nil. In general, he has been forced to link his agriculture with world markets, and has not been able to make the necessary adjustment. Perhaps the core of his problem is the fact that in the majority of cases he is a tenant or a share cropper, barely making a living and possessing no incentive to improve the soil he does not own.

Limited Extension Experience. Extension work, in its modern form and orientation, is still in the formative stage in the Middle East. However, there are unmistakable indications that the various national governments, notably in Egypt, Palestine, and Iraq, have become increasingly conscious of the need in recent years and have embarked upon action programs to improve the lot of their peasant masses.

A general appraisal of the situation reveals the following common features. In each of the countries considered there is a central department or ministry of agriculture. Attached to this are a few or several experiment stations, where scientific research is carried on with a view to improving crops and livestock. Agricultural agents are maintained in the field, but their number is relatively small and so much of their time is taken up by routine administrative duties that they actually can do little for the cultivator. Agricultural literature is available to a limited degree, but the average *fellah* derives negligible benefit from it, because he cannot read. Agricultural education is meager relative to the need of the region. Egypt has four intermediate schools and two colleges of agriculture; Iraq and Syria have one school each; Palestine has one for the Arabs and several for the Jews; and Lebanon and Transjordan have none. So far, home economics work has received little or no attention. Regular school education, in so far as it can serve as a medium for extension work, suffers in two respects: the relatively limited number of elementary schools and the domination of the village school curriculum by city-school standards. It is only in Egypt that compulsory education is prescribed by law, but facilities are still inadequate to make the law very effective.

Within this general picture there are certain achievements or endeavors that need special mention. *Egypt* leads in this respect. Its original effort for agricultural improvement dates back to the middle of the last century. At present the Egyptian Ministry of Agriculture has a special section for extension, whose function is to put available agricultural knowledge in the hands of the cultivators. Recently a comprehensive program has been launched. Its goal is the establishment for every 15,000 acres of cultivated land an Agricultural Center with qualified staff, with whom the *fellahin* could cooperate in the improvement of their crops and livestock. Also the Ministry of Social Affairs has established a special *Fellah* Bureau, which has already started a number of Rural Welfare Centers to help the village people in various ways. The Cooperative Section of the Ministry of Finance sponsors the development of agricultural cooperatives in the villages. At present there are some 1,200 societies which extend credit to the *fellah* and educate him in the improvement of his agriculture and his community.

In *Palestine* an outstanding extension activity among the Arab cultivators has been undertaken through the village school. The project at first consisted of the yearly selection and training in agriculture of a number of village schoolteachers. The second step consisted of application and follow-up work in the village. The teachers returned to their schools and introduced agricultural courses into the curriculum and established school gardens. Students worked in them and farmers came for guidance and to copy from the demonstrational plots. Gradually many of these village schools became dynamic centers of community uplift. The scope of the work was widened to include, besides agriculture, health, recreation, and home welfare.

In *Lebanon* and *Syria* the Institute of Rural Life at the American University of Beirut pioneered in the field of extension work. Supported by the Near East Foundation of New York, it began its work by training village boys in agriculture at its farm school. Follow-up work with these boys was then maintained. Gradually the Institute personnel began working with adult farmers. This led to the organization of short agricultural courses for them and to the publication of pertinent bulletins in simple Arabic. Later on a special two-year course in farm management was organized for the sons of wealthy landlords. Its goal is to induce these heirs to stay on their estates and to interest them in the improvement of their tenants' condition. The Institute also pioneered in the field of agricultural cooperation. So far it has concentrated its effort mainly on one village, where it has helped the people to develop a successful marketing cooperative. A promising volunteer movement, called the Village Welfare Service, has flourished under the sponsorship of the Institute. It began among the faculty and the students of the American University, but in a few years it spread rapidly among the educated youth of Lebanon, Palestine, and Syria. It reaches the *fellahin* directly through camps, where work is done by experts and assistants in health, agriculture, literacy, recreation, and home welfare.

Iraq has embarked in recent years upon a vigorous policy of agricultural improvement. Extension work all over the country is directed by two sections in the Department of Agriculture: 1) Extension and Demonstration and 2) Plant Protection. Field agents are maintained in various localities and their number is increasing. Five experiment stations are now maintained which perform the functions of training skilled agricultural labor, conducting experiments and demonstrations, and disseminating improved seeds and livestock breeds. Agricultural education is being encouraged. There is a school for the training of junior agriculturists, and an increasing number of students has been sent abroad to specialize in various branches of agriculture.

With a view to improving the general extension organization, as described above, the following suggestions are made: 1) In the first place, a bona fide central agency should be established and should be so integrated as to supervise any extension work undertaken by its personnel and by the personnel of other government of private agencies. 2) Provision should be made for the badly needed home demonstration work. 3) The number of field workers should be increased as rapidly as possible, and they should be provided with more adequate and continuous professional training. Also bona fide experience in and sympathy with village life and its problems should be made a prerequisite for training or employment. 4) Finally, experience has shown that the educated youth of the Middle East is willing and eager to volunteer for some form of constructive patriotic service. Here is a tremendous source of energy that should not be neglected in a national program of rural extension.

SUGGESTIONS FOR EFFECTIVE EXTENSION IN THE ARAB VILLAGE: *Where and with Whom to Begin?* An attempt by the field worker at answering this question may be taken as the first step in an extension program. If possible, the initial project or projects should involve the whole community. This would insure stronger and wider support and eliminate possible friction and misunderstanding. At the same time, when the deal is made with the group as a whole, through the council of elders, individual farmers will not fear ostracism and will be more willing to adopt suggested measures. As mentioned above, social control in these compact settlements is strong and effective, and the *fellah* will hesitate to take a new step independently. In facing such situations, individuals manifest their community consciousness through such statements as, "The whole village is for the new idea," "The village is against this innovation," "I like what you suggest, but I cannot stand against my village."

In case the above approach does not seem to be feasible, then a beginning should be made with those individuals or family units who are willing to cooperate and who possess prestige. The teacher, the sheikh, the imam, the priest, a successful farmer, or any other recognized leader are usually accorded the privilege of taking the initiative and setting the example for others. The socioeconomic status of the family or kinship group to which the cooperating individual belongs is also significant in this respect. During his intensive field work in Palestine, Lebanon, and Syria, the writer had occasion to prove the validity of this principle in scores of cases, involving health, agriculture, home work, recreation, and literacy campaigns. Typical responses were: "What is good for Um Ahmed (mother of Ahmed) [1]

[1] According to the local practice of naming a mother or a father after the first son.

is good for me"; "If Abu Hani (father of Hani) is willing to try this new variety of seed, I am willing"; "Who is this so-and-so good-for-nothing that we should follow his example in adopting these innovations?"

Another obvious technique that one should not fail to adopt is the choice of the place best qualified for demonstrational purposes. The community in which work is to be done should be centrally located with respect to surrounding villages and easily reached by them. It should also be a fair representative of the prevailing conditions and problems. It is in this respect that the cooperative movement in Egypt and Palestine has been to some extent criticized. It is pointed out that agricultural cooperatives have been a success partly because they have been established in "good" villages to begin with, whereas the communities that need societies most have been neglected. The same qualifications apply equally to the choice of the demonstrational center within the community. For example, agricultural demonstration would be most effective if undertaken on plots next to the school, the mosque, or the church. Another good location is the center of a stretch of open fields planted to the same crop by the various owners. A village teacher in Palestine made effective use of this point. He convinced one farmer to spread several loads of farmyard manure on his field before planting it to wheat. He wanted the farmers to realize the value of manure and to desist from the practice of heaping it within the village or burning it. The difference in growth and later in yield between the manured wheat and the surrounding plots was the most convincing sight for the village to behold for the whole of that season and for the following seasons too.

For various reasons, there are in any locality some villages that are relatively receptive to new ideas and others that are resistant to change. It should not prove difficult for a worker to detect such general characteristics of communities. Other things being equal, the contemplated program should be started among those groups who are more tolerant of receiving strangers and accepting new techniques. The Lebanon villages, for example, are in general of the tolerant type. This has been partly due to intensive emigration and contact with Western educational institutions. After a short period of field experience, and in cooperation with local government officials in Palestine and Syria, the writer was able to classify communities as "easy," "difficult," "very difficult," and, sometimes, "hopeless"!

Another observation to be made in this respect is that Projects are carried to completion more quickly and more effectively in settlements already suffering from certain problems. The people affected are eager to find a solution and are willing to offer the necessary support. This point will be discussed below from another angle.

Get Acquainted with the Local Situation. This step may be taken first or second, depending upon how much the worker already knows about the locality. In the first place, this entails a general knowledge of the culture and its main pillars or foundations and values. Such knowledge serves as a general background upon which the field worker can draw as the occasion demands. For example, without such general orientation an outsider may propose the raising of pigs in a Moslem village. From an economic point of view it may be a sound project. The reaction of the people, however, is certain to be strongly negative, because of the religious taboo imposed upon these animals. Not only that, but a prejudice against the worker and his subsequent projects may result.

Another illustration is afforded by an incident which the writer and a colleague of his had in an outlying village on the borders of Transjordan and Palestine. Because infant mortality was very high and the general health situation bad, a sanitation project was being considered. During the discussion with the village elders the colleague made a remark which almost turned the group against the proposal, although what he was saying was correct and scientific. He referred to the many small tombs at the village entrance, indicating a high infant mortality, and emphasized the ability of science to rectify the situation. The imam (Moslem religious leader) was quick in his reaction. He said that that was blasphemy, that the death and the life of a child were in the hands of Allah and that human effort was of no avail. The rest of the elders readily murmured agreement. The situation was finally rescued by the teacher who recited from the Koran to the effect that man should always do whatever little he can and depend upon Allah for the rest.

In addition to gaining knowledge of the general pattern of culture, one should make a study of the socioeconomic organization of the particular community concerned. As mentioned above, practically every Arab village has its own character and distinguishing features. An authentic knowledge of these can be gained only by actually living in the village and participating in its affairs. Obviously, the way of life in a Druze settlement is not the same as that prevailing among Moslem and Christian groups. Likewise, the Kurdish village in the north of Iraq differs in many respects from the settled tribe in the south, although both are Moslem. The local situation also differs markedly according to whether two or more sects live in close proximity or are segregated. Further, each group has its own historical background, which is influential in the present organization and behavior of its members. There are also important local differences with respect to type of agriculture, level of income, feuds and conflicts, cooperative spirit, type and number of leaders, and values that are specifically emphasized. The more the worker comes to know these intimately and in detail the better qualified he is for his extension activities.

In this appraisal of the local situation special attention should be paid to community problems and needs, the solution of which is the concern of extension work. After compiling an exhaustive list of these, the worker should classify them according to three points of reference: urgency, complexity, and whether they are felt by the people or by the worker himself. This classification will help him in deciding on the initial steps of his program.

In some communities it may be discovered that certain new ideas and practices had already been introduced and established and others had been rejected. A study of such cases should be helpful in revealing the means by which success was achieved or failure resulted.

What Projects Should Be Undertaken? No categorical answer to this question can be made. One has to appraise the total situation and decide each case on its merits. With this qualification in mind, a few generalizations may be attempted. Quicker and more permanent results may be expected if the program is started with projects that meet the felt needs or problems of the community. The soundness of an extension activity can be measured by the degree to which it makes people progressive-minded and self-helping. In such cases active local support is likely to be given and permanent interest in the project maintained.

In the village of North Lebanon where the writer and a group of volunteers were doing some extension work, the people responded readily to those health projects that involved curative medicine. It was a need they felt keenly. They wanted to be rid of disease. They also accepted agricultural measures aimed at saving their crops from pests or increasing their income. It was a completely different story, however, in the case of an attempted economic survey of the community involving questions about individual incomes. Not only did the villagers consider the questions foolish, but they also began to suspect these "strangers" of ulterior motives, including the possibility of helping the government to increase their taxes.

The program should begin with the more simple projects that can be finished in a relatively short time and that show tangible results. Complicated, large-scale problems should be attacked later after the people have experienced the fruit of concerted effort.

A third suggestion is that projects of interest to the whole community have a better chance of success and make a more auspicious beginning for an extension program than those that concern one or a few individuals. Practically every village offers possibilities of this kind. Common examples are sanitary water supply, road construction and repair, community playground or center, planting of shade trees, irrigation schemes and agricultural demonstration on *waqf* property.

Finally, it is safer to start with local issues that are not heavily charged with emotional content. In the course of a general survey the writer was conducting in a village, things ran smoothly until he began to question girls about the kind of home they wanted to have in the future. An explosion followed. The question implied marriage, a matter about which girls do not talk in public. In another case, a village teacher, who was eager to start a garden for his students, took them out one day and began clearing a waste plot at one side of the community cemetery. The disturbing news spread around rapidly that the graves of their ancestors were being desecrated. The atmosphere became tense with aroused emotion, and the teacher had to relinquish his project.

Specific Techniques for Carrying Out Projects. Perhaps the most effective technique that can be used to arouse the *fellah* for appropriate action is the *emotional* appeal. It has been shown above that he is demonstrative and that he derives satisfaction from indulgence in such demonstration. This is a delicate approach, and one should not attempt it unless he is thoroughly acquainted with local folkways and cultural values and is able to use the right words and phrases at the right moment. Some of these appeals or factors of motivation are general and apply to all Arab communities, whereas others vary from locality to locality. The main ones among these will now be discussed briefly.

Citations from the Koran. This is the sacred book and the source of a total way of life for the Moslem *fellah.* In most cases he does not know how to read it, but he has heard and memorized many of its verses. A quotation is sometimes the most convincing argument a field worker can use in support of a new idea or a project. The manner in which the citation is invariably introduced indicates its final authority. "Allah hath said in His exalted book. . . ." It is literally Allah's word. Such statements of general appeal as "O, followers of the prophet! . . ."; "All those who love the prophet! . . ."; and "He who is willing among you to honor his religion and Allah's book!" seldom fail to arouse active response. At the same time there is a wealth of Koranic statements pertinent to improvement in agriculture, health, cooperation, literacy, and other aspects of life.

Personal, Family, and Community Honor. The Arabic word for honor is *sharaf;* it is a widely used and highly significant concept. In his early childhood an individual begins to be impressed with its importance. He learns to live up to and defend his *sharaf* and that of his family and community. "He lacks honor!" "He belongs to an honorable family!" "The honor of our village is at stake!" These and similar expressions are often heard in a variety of situations where discussion and argument over personal or group issues are involved. Successful appeal can be made to a simple *fellah,* to a leader, or to a whole community through this sensitive point of contact. In a project of village beautification, which naturally involves the desolate cemetery, motivation for action can be achieved through reference to the "honor" of the dead; and the *sharaf* of the whole group can be readily enlisted in the effort to establish a credit society and get rid of the exploiting usurers.

The Glorious Past. In some areas the *fellahin* know a good deal about Arab folklore and past civilization; in others they know very little, and one has to educate them. Showing lantern slides of Arab monuments still standing in various parts of the Middle East, lecturing on the Arab contribution to world civilization, and giving the people a graphic description of the successful agriculture their ancestors developed in Iraq, Syria, Egypt, and elsewhere inevitably brings more favorable results. Finally, the *fellah* of today must be told that he is not inherently inferior to his progenitors and that he can emulate their example if he wishes to do so.

Competition. This may be employed constructively on the inter-group level, between families or parties within the community or between communities. In one village two societies tried for several years to excel each other in play production and in welfare activities. In most cases it was found that when one family adopted a new agricultural technique successfully, other families felt challenged to try it, too. The motive of rivalry came to the foreground when the American University of Beirut established its first extension camp in a village of northern Lebanon. By the end of the summer (extension season) two neighboring villages appealed to the director to transfer the camp to their own territories. They felt better qualified to manage it than the other community, and they were willing to cooperate in all projects suggested by the expert. "If the people of 'Aba can build a school, we can build a better one!" "Here are our neighbors in Ramah getting so much more oil from their olives, because they do the picking and the pressing according to the new methods of the Department of Agriculture. We certainly can adopt these methods and do as well as they can!" "From now on it is going to be either Beitunia or Saffa; the whole country shall know which one will produce more and better grapes!" Such are a few examples of the way the spirit of intervillage rivalry is expressed. There is, of course, a danger that competition may reach the extreme of conflict and feuds, and result in the disorganization of community life.

The Spirit of Nationalism. This is rather a new factor which has emerged on the rural horizon in recent years. Not longer than three decades ago, the village was the *fellah's* cosmos, beyond whose limits the radius of his life's activities seldom reached. His three major loyalties were for his family, his church, and his community. Under the impact of the newly awakened spirit of nationalism, this isolation of the *fellah* from the affairs of the country is coming to an end. As a cultivator of the cotton fields in Egypt, or a tiller of the soil in Palestine, or a partly settled tribesman of Iraq, or a Druze mountaineer, he has become awakened, in one degree or another, to a consciousness of the identity of his destiny with that of his country. Here is a tremendous driving force, which has been exploited in the struggle for political emancipation by these countries and which can be similarly directed towards the reconstruction of their rural life. It is not far-fetched to appeal to the *fellah* to use a better plow or better seed, the landlord to accord his *fellahin* better conditions, and the educated youth to volunteer his services for village folks on the basis of patriotic duty. It was mainly on the strength of this motive that educated Moslem girls from Damascus took off the traditional veil of centuries and joined the camps of the young Village Welfare Movement.[2]

In addition to the emotional appeals listed above, another suggestion for carrying through the project is that it should be preceded by a preliminary educational campaign. This may not be necessary when the people are aware of the problem and feel the need for a solution. In many cases, however, it is the worker and not the people who sees the difficulty. In such cases his task is doubly difficult. He has to proceed slowly, first educating them with respect to the existing problem, and then getting them to support the necessary project. In view of the prevailing illiteracy of the *fellah*, it has been proved that the visual type of education is the most effective. Pamphlets are of little value. If posters are used they should be very simple and direct. Moving pictures are greatly appreciated, but much of their instructional value is lost because they move too fast. Still projections are superior in this respect. The ideal is a combination of both, a moving film projector that can be stopped at any point for a reasonable length of time. For a period of two years the writer made intensive use of such a projector and proved its effectiveness. The radio has been tried successfully in some villages and it promises to become a potent educative force among the illiterate *fellahin*.

The following incident is given as an illustration of the need for an initial educational campaign. In one village, where trachoma incidence ran exceptionally high, the medical officer of the district and the writer launched an eye-treatment project through the local schoolteacher. They expected the villagers to report to the school readily and submit to the simple daily treatment willingly. They felt that the whole group was conscious of this ailment, and that everyone who was afflicted was anxious to get rid of it. They soon realized, however, that they were taking too much for granted. The people were more or less indifferent in their response. They had had trachoma for generations and accepted it as normal. One day, in the course of a general

[2] For an analysis of this young rural movement see A. I. Tannous, "Rural Problems and Village Welfare in the Middle East," *Rural Sociology,* VIII, No. 3 (Sept., 1943).

meeting, the writer asked one of the leading elders what he thought of the health situation in the village. He felt it was fine. The medical officer pointed to the swollen eyelids of most of those who were present. The elder did not see anything wrong there. Finally, when a man who was completely blind in one eye was brought forward, the general remark of the audience was "This is from Allah." At that point the promoters of the project were convinced that they had to start from the very beginning and show the community that there was something wrong before attempting a cure.

A third suggestion is that the worker should get the cooperation of as many people as possible before they have time to go on record as opposed to the suggested line of action. Almost invariably there are individuals in each village who wait for a chance to express themselves negatively, especially if they have not been consulted about the matter. Once they take a public stand to that effect they are more or less committed to it. In more than one village our whole program was seriously hampered because of opposition which, with a little more tact, could have been averted at the start.

A fourth technique is to make use of existing village organization. For example, a new building is not indispensable for a community center; the school, the mosque, the church or the *madafs* (guest houses) of the various kinship groups may serve the purpose. A credit cooperative society may be established on the basis of the local system of *kafalah mutasalsilah*, whereby unlimited responsibility is assumed by the group for the loan they secure from the bank. In each community there is either a village or tribal council, or a municipality. Projects may be entrusted to these agencies. The village school in Palestine has been used effectively as an extension center. The teachers and students have been able to help in the development of cooperatives, first aid treatment, sanitation, agricultural demonstration, and literacy campaigns. Local Boy Scout activity may be easily directed towards village improvement. Many communities have literary societies which may very well sponsor an adult education project.

A fifth point, which is closely related to the one just mentioned, is that a program should be formulated on the basis of what the people already know and have. This is a well-known educational principle, but it is still being neglected in many instances. The promoter of an idea or a line of action is naturally inclined to use his own background rather than the local situation as a base of operation. The *fellah's* wooden plow, for example, has been readily condemned by many "experts" as a primitive and ineffective tool, and its replacement by the modern plow recommended. Upon more careful observation in the field, it has been found that the *fellah* is not inclined to desert his traditional tool, which actually is well adapted to the local situation. A wiser course would be to modify the iron tip, instead of discarding the plow completely. An enthusiastic specialist may attempt to have the *fellahin* build modern poultry houses and supply them with new breeds. Sooner or later he comes up against obstinate difficulties, for such houses are beyond the *fellah's* financial means, and his standard of sanitation is below what the "foreign" chickens require to keep in good condition. A more effective approach would be to help the *fellah* select the best specimens in his flock, and thus improve its quality gradually. The people of many villages gather and eat certain wild herbs which are rich in valuable minerals and vitamins. A home worker may ignore this and similar "exotic" food practices in favor of her own standard, or she may take them as a basis for the development of a diet that is best suited to local needs and conditions.

Utilization of local leadership is another principle that the field worker should keep in mind. It has been shown above, that each village has its own leaders. One or two of these may be key leaders for the whole community, whereas others assume leadership in specified situations. In addition, each kinship group has one or two leading elders. Much better results can be secured if projects are undertaken with the consent and support of these individuals. If the imam consents to have the mosque used for a lecture meeting, the orthodox villagers will be less likely to object. If a leading elder can be persuaded to have his daughter take a practical nursing course, other families will follow suit. A new agricultural practice will be adopted more easily if it is sponsored by a recognized farmer. In many instances the worker should be careful not to mistake apparent for actual leadership. Even in these little villages there may be a "power behind the throne."

Finally, let the enthusiastic extension agent not forget that his ideal goal is "of the people, by the people, and for the people." He should remember that he is an outsider to the community and that tomorrow he may not be working in it. Consequently, he should be constantly on his guard against the common and tempting pitfall of building the program around himself as an indispensable center. It may be easier for him to do so, he may achieve quicker results, and he may have good reports for his superiors. In the long run, however, both he and the community will be the losers. An extension program that is directed and supported mainly from above and from without will not have much chance of becoming an organic and permanent part of community life. A sound procedure in this respect is to give the people the chance to talk the project over and develop public opinion about it, assume responsibility for directing it, and defray its expenses in labor, in kind or in cash.

Here is a pertinent instance. A wealthy emigrant returned to his native village in Lebanon for a short visit. In good faith, he wanted to make a worth-while contribution to his community. He conceived the idea of a school, without stopping to consider that there must be reasons why the village had not previously had one. He contributed all the necessary funds, and a modern building was erected. He organized a committee to take charge in his absence, and promised to pay the teachers' salaries, so that all children might have free education. Then he left. The school ran for two years, during which many inevitable local factors, which hitherto had been ignored, asserted themselves. The committee was split against itself, bickerings developed, and funds were abused. In the third year nothing was left of the project except the empty and neglected building.

THE VILLAGE PUMP. Such, then, is a brief picture of the community organization in the Arab village, and of the possibilities, difficulties, and techniques involved in rendering that organization more effective, which should be the ultimate goal of any form of extension, relief, or rehabilitation. The detailed story of the way one project was fitted into the organization of an Arab village should serve as a fitting conclusion. The story is rather typical, and since it occurred in the writer's field experience it is told as a personal narrative.

One of our Village Welfare Camps was established at the main spring, just outside Jibrail, a foothill village in the extreme north of Lebanon, where most of our work was centered. About two miles away stood Ilat, a small community of a few hundred people. One morning a few of them came to the camp and asked for medical help, saying that many of their children were stricken with "fever," the word they used to cover all sorts of internal diseases. Our doctor and two assistants went to the village to investigate. They came back in the evening and reported several cases of typhoid, malaria, and dysentery, and a high incidence of infant mortality, that was another case of the general health problem common to most villages. We promised to extend to them medical treatment and to do what we could along preventive lines, which was our general policy in solving the problem.

Further investigation revealed the probable source of trouble: a tiny spring in the midst of the village, which flowed into a stagnant pool. It was the only source of water supply, and we saw how it was being utilized. One woman after another emerged from the surrounding houses, each carrying an empty jar in her hand. (Hauling water is a woman's job, and a man would be ridiculed if he should be seen doing it.) With bare feet they walked in the dirty street, waded into the pool, drank, and gave their trailing children to drink, filled their jars, raised them to their shoulders, and walked back home. Animals came to the pool too—cows and oxen, goats and sheep, and donkeys. They waded and they drank. So we thought that our line of action was clear and simple. Dig the pool deeper, cover it with a stone structure, and install a hand pump. It was as simple as that.

One evening we called the elders to a meeting and informed them of our plan, requesting them to render as much help as they could. There seemed to be general agreement. In our lack of experience, however, we had not yet learned the subtleties by which a "yes" may mean a "no" in certain cases. The following morning, when we came to the village, ready to begin the project, we found the place practically deserted. They had all gone to their fields. The *mukhtar* (Headman, a government official) made his appearance to tell us that the people refused flatly to let us install the pump. *Let us install the pump!* That made us pause and think. So that was how they felt about it; that we were imposing on them something they did not really want. And all the time we took it for granted that we were satisfying their urgent need. Something was certainly wrong.

With much difficulty we were able to bring them to another meeting a few days later. In the course of the discussion we did our best to make them talk freely; and they told us a great deal. The following are more or less direct quotations:

"Our fathers, grandfathers, and great grandfathers drank from this water as it is, and I don't see why we should change now."

"*You* say that *you* want to install a pump at the spring; but I for one have never seen a pump, nor do I know what might happen if it should be put there."

"I tell you what will happen. The water will flow out so fast that the spring will dry up in no time."

"Not only that, but the iron pipe will spoil the taste of the water for us and for our animals."

"You So-and-So," put in one of Jibrail's elders, who are much more advanced in their outlook than the people of Ilat, "do you like the taste of dung in your water better?"

"Well, I admit it is bad; but we and our animals are at least used to it."

"You have told us that the water is the cause of our illness and of our children's deaths. I do not believe that, and I can't see how it could be. To tell you the truth, I believe that the matter of life and death is in Allah's hands, and we cannot do much about it."

"One more thing. We don't understand why you should go to all this trouble. Why are you so concerned about us?"

"You say that the pump will save our women much effort and time. If that happens, what are they going to do with themselves all day long?"

At the close of the meeting we realized that we had blundered. We had to begin from the beginning, taking nothing for granted. An educational campaign was launched, starting with laboratory tests of Ilat's water and samples from neighboring villages. We emphasized to the people that the report on their water was *very bad*, whereas the other villages received *good* reports. The way the hand pump worked was demonstrated to them, and they were convinced that it would neither spoil the spring nor dry it up. Quotations from the Koran were cited to the effect that cleanliness was required from every faithful Moslem and that man should do his best to avoid the danger of disease. At the same time, our girl workers visited with the housewives and explained to them how the pump would make their day's work easier and how they could use the time saved in taking better care of their children. They would not get ill so often, and fewer of them would die. Finally, we did our best to explain to the villagers that we were doing all this as our patriotic duty, and that it was their duty also to cooperate with us for their own benefit.

It took one whole month before the situation was ripe for action.

We advanced the cost of the pump and its accessories, which we ordered from the neighboring town. We insisted, however, according to our working principle, that they should contribute the necessary labor and pay in cash or in kind as much as they could. Two of our volunteers took with them a donkey and went from house to house gathering contributions. Towards the evening they came back with a small sum of money and a heavy load consisting of barley, wheat, eggs, chickens, and fruit. The following morning the villagers started working. The pond was cleaned and deepened; a stone structure was built over it, and the village pump was installed at last.

Chapter 8 · EXTENSION IN THE BALKANS · By Clayton E. Whipple

ECONOMIC BACKGROUND. The Balkans comprise six countries—Hungary, Rumania, Yugoslavia, Bulgaria, Greece and Albania—with a total area of 346,000 square miles (about a third larger than Texas) and a total population of about 59 million. More than two thirds of the population of the region consists of agricultural people, living in small villages and owning and tilling, on the average, about 12 acres of land in small strips of less than an acre each scattered about the community. Hungary is an exception to the rule, since the prevailing system of land tenure in that country is the *latifundia* or large estate. Vestiges of a similar feudal system persist to some extent in parts of Rumania and Albania. Land reform in Rumania and Yugoslavia was effected to a considerable extent after the last war and many peasant families were established on small farms.

In all Balkan countries the pressure of population on the land is heavy.[1] Careful analysis of official statistics for the period just preceding the present war indicates that about 63 per cent of the total surface of Hungary was suitable for agricultural use in some form, while corresponding percentages for other countries in the Balkans indicated 50 per cent in Yuoslavia, 45 per cent in Rumania, 40 per cent in Bulgaria, but only 22 per cent in Greece and about 15 per cent in Albania. Hungary had 24 agricultural people per 100 acres of total agricultural land, Rumania 30, Bulgaria 33, Yugoslavia 42 and Greece about 70. Authorities agree that about 3 acres of land are required to maintain one individual in terms of the limited diet and few material comforts traditionally secured by the peasant families of the Balkan region. Further analysis of the information available shows that per head of farm population Hungary has 3 acres of arable land, Rumania 2.4, Bulgaria 2.2, Yugoslavia 1.8, and Greece only 1.3, while the small amount of arable land per individual in Albania is indicated by the fact that the country is unable to feed itself adequately despite the fact that 85–90 per cent of the population is agricultural. All of the above information may be summarized and condensed into a statement

---

[1] For a more complete discussion of this topic consult C. E. Whipple and A. U. Toteff, "The Problem of Surplus Agricultural Population in Peasant Countries," *Journal of Agrarian Affairs*, I, No. 1 (1939), 61–78.

that the average peasant family in the region consists of 5 or 6 members, living in a small one-story house, with a garden enclosed by a mud-brick wall, and farming about 12 acres of land divided into 10–15 small strips scattered about the community. Each family also has a share in the community pasture and a large share of the feed for the one or two draft animals and (or) several sheep and pigs is secured from the wild grasses growing on this community pasure. The family also has a share in the community wood lot, if one exists; and if not, burns straw or dried manure for fuel, thus further reducing the fertility of the fields already depleted through generations of cereal farming and little or no use of artificial fertilizers or soil conservation practices, apart from an occasional year in fallow.

Paradoxically, extensive farming, stressing the production of wheat and corn, with some nomadic herding of sheep and goats, prevails throughout most of the region. A more intensive agriculture would more effectively utilize the large potential labor supply indicated by the figures cited above. In fact, over 75 per cent of the arable land of the region is planted to cereals, which produce a low income per farm and per acre; logically enough, cereals form over 50 per cent of the caloric intake. Cash income is low, averaging only $50 to $75 per farm in the region and—apart from a limited number of farms in certain regions of Greece, Bulgaria, and Yugoslavia, producing and marketing tobacco, raisins, or olive oil, and therefore buying most or all of the food consumed—subsistence agriculture is the prevailing type. It is true that the majority of the meager amounts of milk and eggs produced must be sold in order to pay taxes, buy salt, kerosene, and other necessities, and this often applies to meat and—where markets are close at hand—to fruit and vegetables. As a result, the diet is limited, protective foods are consumed in small quantities, food preparation is inadequate, and facilities are not available generally for preserving temporary surpluses of perishable food.

A soundly developed system of land reform in Hungary and in the sections of Rumania and Albania where large estates still exist, would assist, at least indirectly, in solving the problem of agricultural overpopulation. A more effective measure for the region as a whole would be the completion of the process of "commassation," that is, regrouping of the scattered strips belonging to the peasants into solid blocks of land. This process has been carried out in a number of Balkan peasant communities and has universally resulted in improved agricultural techniques such as better tillage, introduction of crop rotations and specialized crops, soil improvement through better use of fertilizer and growing of leguminous crops and more efficient use of labor and draft animals—all resulting in better yields and higher farm incomes.

Previously, emigration was an important factor in reducing rural overpopulation, but in recent years political and economic regulations have sharply reduced emigration from Balkan countries. Reopening of emigration opportunities to any significant extent in future years appears doubtful.

Much attention is now being given to development of industry in the region through measures varying from a gigantic Danube Valley Authority to less ambitious and more local developments. At present, some small-scale industry exists in all of the countries involved; it seems logical that more capital and labor will be utilized in small-scale industry, particularly in processing agricultural products and making articles for peasant households. Sericulture and preservation of foodstuffs would employ some people full time and many part time, and the latter would prevent wastage and improve the diet, while the former would provide additional income. Some additional development in larger-scale industry may result, but probably not enough to employ many additional thousands of the surplus agricultural population. The countries lack sufficient funds, and also financial organization as well as managerial and technical skill, to permit of successful competition with the West. Industrial wages are low and the resulting standard of living is often on the one hand lower than that of the peasant population and more uncertain on the other. The peasant at least has shelter and food and, in fact, lives best when markets are temporarily unavailable or prices too low to encourage much marketing.

Hence the solution which would assist the most people, in fact a majority of the total population of the region, would be the reorganization and improvement of the small peasant farms from which the majority of the population secures its living. Careful studies of the results of improved farm practices, with resulting increased income, in demonstration villages in Greece, Bulgaria, and other Balkan countries have shown that under proper guidance the increased cash or barter income is utilized for the purchase of products of local and, to a lesser extent, of foreign industry. This includes such items as dishes, kitchen equipment, shoes, clothing and other household items previously lacking, as well as better plows and harrows, grain drills and other farm machinery.

In these rural communities the average cash income was increased from $40 to $50 annually to as much as from $200 to $300, and in individual cases even more cash was available. This resulted in the development of local and regional industry on a small scale and even to importation of machinery. Unless the millions of peasant families secure increased purchasing power there will be little market outlet for industrial products. Moreover, existing industry will continue to operate on such a limited scale that wages will remain at existing unsatisfactory standards and neither the peasant nor the worker will have the means of purchasing the products of the other.

SOCIAL BACKGROUND. In view of a thorough and careful analysis of peasant society elsewhere in the book,[2] it seems unnecessary to describe at length the socioeconomic background of the area. The rural people of the Balkans live in folk societies which have stood the test of time. The influence of kinspeople is still very great, though the individual family has gradually superseded the old patriarchal or joint family system, the "Zadruga" system of the Slavs. The family is the center of village life, the basic economic as well as social unit. It is the producing and consuming unit, the determinant and perpetuator of customs, beliefs, traditions, and values. The head of the family is still an autocrat in his family, the more outstanding family heads are the village "elders" and in most cases are still more important and influential in determining community attitudes and carrying out programs than are officials, transplanted from outside by national authorities. Generally the mayor is a city man, usually with a legal education. The priest and the schoolteacher are often local men and, if so, have more than usual influence through a combination of the prestige traditionally enjoyed in the Balkans by the educated class and of knowledge of how things are done in the village. The role of the doctor, agricultural specialist, and veterinarian will be discussed under extension methods and personnel.

The Balkan peasantry are hard working, frugal, hospitable, able and willing to learn, and friendly to all who come with sympathetic understanding of their problems. Every peasant community has one or more men and women who are locally outstanding and considered

[2] See Chapter V.

leaders by their fellow peasants. Officials working informally with these influential local people have developed and carried to a successful conclusion outstanding programs based upon problems considered important by the group and developed within the framework of the capacity of the local community to carry them out.

The enduring conservatism of the Balkan peasantry does not stem entirely from rigid adherence to tradition, but is also perpetuated by the prevailing meager resources in capital, land, equipment, and scientific knowledge. It has protected him from embarking on many unsound adventures in "improved" agriculture, not suited to his conditions. This conservatism is, therefore, an asset rather than a liability to the extension worker who, in cooperation with the local people, develops a sound program which fits local conditions and which can be carried out cooperatively in accordance with local customs and with local resources.

COOPERATIVES. One of the most valuable assets available within the region is the general-purpose type of cooperative found in the majority of Bulgarian rural communities and in many rural communities in the other Balkan countries. A properly organized and operated cooperative enables the members not only to buy, sell, borrow, and insure in larger and more economic units, but also to own and operate cooperatively grain drills, seed cleaning and disinfecting equipment, threshing machines, incubators, spraying machinery, butter- and cheesemaking equipment, and other production and marketing essentials. In many communities the cooperative either owns or provides capital for the animal husbandry breeding associations to purchase high-quality bulls, stallions, rams, and boars for improvement of the local animals. The cooperative is a most valuable asset in any program of village improvement, as it utilizes the time-honored method of working together and provides the necessary production factors and a more efficient and profitable means of marketing the surplus products.

RESEARCH AND EDUCATION. Agricultural research is being conducted on a considerable scale in all of the Balkan countries. The quality and nature of the work varies from country to country and station to station. In some countries research in plant improvement and soil preparation has surpassed that in animal improvement; one or two countries have done outstanding work in plant pathology but little in entomology, while in some cases the most outstanding scientific work has been done in horticulture. Generally speaking, research work in agricultural economics, and particularly in farm management, has not kept pace with research in plant and animal science, though there are notable exceptions in one or two countries. Unfortunately, too little of the results of the research has been carried in an acceptable manner to the peasantry of the region. Outstanding work in public health has been carried out in all of the countries during the present generation, and doctors and nurses have made outstanding application of the principles of public health extension in some rural communities in several of the countries. Unfortunately, little research has been done in rural home improvement. Each country has several stations and usually two or more sections exist in each station; but a unified, integrated national program of research is lacking to at least some extent in all of the Balkan countries, though progress is being made in such integration.

Elementary schools exist throughout the region apart from a limited number of villages in one or two countries. It is seldom necessary to enforce compulsory educational laws. In fact, some peasant families make great sacrifices in order to send an outstanding son to the gymnasium in a near-by city. Graduation with a high average grade from a gymnasium (roughly equivalent to completing our junior college) is necessary in order to enter the university. Hence peasant youth, especially girls, comprises only a very small percentage of the student body, and as a result few doctors, veterinarians or agricultural scientists, extension agents and teachers come from peasant families, and few peasant girls go to the homemaking normal schools. The theoretical nature of the instruction in the universities and the lack of farm practice handicap the prospective rural worker, who usually does not understand rural customs and traditions and does not know practical agriculture.

Higher education in the Balkans has tended to stress the preparation of the few for the professions—law, medicine, architecture, literature, and business. Even rural elementary education is of the classical type, and the gymnasium is almost entirely so, despite some attempts at reform. More emphasis must be placed upon mass education of a simple, direct type, taught in close proximity to the farms and homes of the peasant millions. In order to do this, personnel with a sympathetic understanding of Balkan rural conditions is required for bringing the results of scientific research to the farms and farm homes.

A soundly developed and operated extension system is essential in every Balkan country in order to raise the standard of living. For the purposes of this discussion extension is defined as informal education which makes scientific knowledge available to and usable by the people who live and work on the land. The desired end product is better living for the rural people, and hence extension must stress home welfare, improved health conditions, and recreational features as well as improved agriculture. Increased farm income and larger and more varied production of food for home consumption must be converted into a higher and more satisfactory standard of living.

Extension, as defined above, differs from the prevailing agricultural services authorized by legislation in most Balkan countries in that it is an educational process carried out cooperatively with the people rather than being superimposed upon them through mandatory regulations and operating devices. Because of increasing population pressure and the penetration of Western civilization, the traditional folk practices are slowly changing. The role of extension in the area, therefore, is to accelerate and facilitate this change, as desired by the people themselves, by bringing to them the best knowledge of science in ways which are practical and acceptable to their cultural backgrounds.

EXISTING EXTENSION SYSTEMS. Every Balkan country has an operating extension system and extension work is being actively conducted in each country chiefly under governmental auspices, though some of it is conducted through cooperatives and other private or semiprivate organizations. Every country likewise has some outstanding achievements to its credit in agriculture, homemaking, and health. But these achievements have been localized or else limited in scope. In some cases an all-round extension program has produced beneficial results in a small group of villages or in a single village. Very favorable results have been secured in some countries over a wide area in one phase of extension work such as disinfection of wheat seed; preventing the ravages of an insect pest of great economic importance; organizing the efficient marketing of an important commodity such as tobacco, currants or table grapes; efficient and economic utilization of all the threshing machinery of the country in all the principal grain-producing regions; effective malarial control in infected regions. In most countries, personnel assigned to extension work is relatively numerous and many of the extension workers have outstanding achievements to their credit.

In all of the Balkan countries extension work in agriculture, veterinary, medicine, and home economics is administered and carried out under the authority of the Ministry of Agriculture, while public health work is under the State Health Department. Unfortunately, neither agriculture nor health has established a specific agency for extension work in any of the countries involved. Each ministry of agriculture has agencies corresponding to the bureaus of plant industry and animal industry of the United States Department of Agriculture and, in addition, a Bureau of Veterinary Medicine. The latter in each country actively competes with the Bureau of Animal Industry for the control of animal research stations, organization and operation of animal breeding stations and organizations and in extension work, while in some countries both bureaus compete with the Bureau of Plant Industry for control of forage crop and pasture improvement research and extension programs. In each bureau one of the senior officials is usually placed in charge of extension work operating under that bureau, but without exception he has many administrative duties, chiefly of a regulatory nature, in addition to his extension responsibilities. One or more junior personnel and some clerical personnel assist him on a part-time basis.

The bureau extension administrators lack sufficient funds for adequate supervision in the field, and in many cases have inadequate telephone facilities. Hence most of the contact must be maintained through reports and written instructions. Until recently promotion was theoretically based on a rigorous application of the principle of seniority, though in practice the frequent changes in political control usually resulted in widespread personnel shifts, made possible through the absence of a civil service system. Only two extension administrators in the region have had professional training in extension work in the United States and the majority of the administrators have had no preprofessional or in-service training in extension, rural education, rural sociology, or even farm management. Many lack a firsthand knowledge of practical agriculture. In the middle 1930s, younger and better trained men were promoted to higher posts in some Balkan countries, and a program involving improved administrative and supervisory techniques began with the assistance of American advisors. This, however, was interrupted at an early stage by the present war. No woman has yet achieved a high administrative post in homemaking extension, and even the homemaking normal schools and nurses' training institutions are directed by men, though the technical training is conducted by women.

Each country is divided administratively into several provinces and each province into a number of counties. Each province has a regional director of agriculture, selected from the Plant Industry Bureau, with an associate or usually an assistant for animal husbandry. A provincial director of veterinary medicine operates independently, and often the two directors are in conflict. In some provinces, specialists in subject-matter fields such as horticulture, plant protection, and tobacco growing assist the director. In Bulgaria, each province has an assistant director who devotes most of his time to supervising the work of the courses in agriculture and homemaking attached to the village schools, and the resulting extension work conducted in the region of the school by the school personnel under the direct supervision of the county agent. Usually, the specialists are merely competent young research men, lacking both a knowledge of practical agriculture and of the technique of assisting the county agents in conducting practical campaigns in the subject-matter field of the specialist.

In some Balkan countries there is a county agent for each county, who sometimes has an assistant specializing in animal husbandry or horticulture, in others one man must attempt to serve several counties. In some countries, each county also has a county veterinarian, and the degree of cooperation or competition between the two men is usually determined on a personality basis. The county agent travels with a carriage, on horseback, or sometimes on foot, usually has an area larger than the average United States county, and must devote a majority of his time to regulatory activities and determining and applying penalties and rewards, rather than to extension work of the educational type. He is usually a well-trained generalist in agricultural subject matter but lacks preprofessional or in-service extension training. Active county agents in the Balkans must be away from home and office for several days or even weeks at a time due to primitive traveling conditions and hence lose touch with their offices. On the other hand, county agents less inclined to travel can usually justify remaining at home and satisfy their superiors by submitting statistics and reports of work done locally through pressure exerted through the mayor, the cooperative, or other agency.

Observations of the activity of many county agents in various Balkan countries has convinced the author that almost without exception the county agent carries out his work with the aid of enforcement devices. He makes his annual programs, utilizing national or provincial instructions and directives and without consulting competent and progressive local peasant leaders. The program is often set in motion by sending an instruction to the local authorities, the more active agents usually visit each community and lecture all the peasants who can be assembled and then turn the program over to the mayor, the leader of the cooperative, if one exists, and other local officials. The county doctor similarly places the responsibility of enforcement upon local medical officers or, if they are lacking, advises the mayor to press for the appointment of one. All these officials are appointed and paid by the national government and all are frequently shifted from county to county.

A study of the programs of several agents in a given country discloses the fact that they are much alike in that they cover many crop and livestock enterprises and give little emphasis to the more important ones. Climate, topography, and cultural factors differ widely in each Balkan country, and often there are several types of farming areas in a single county. Usually the county agent has over a hundred items in his program of work, many of them of very small importance in a given county. Concentration on a few problems of major importance has stimulated peasant interest and cooperation in many counties. This has provided further benefits in that the county agent has found his work more satisfying and rewarding because of local approval, and has been stimulated to remain longer in a given county. Better human relations and a tendency to consult outstanding peasants in program preparation and development usually accompanies a longer tenure. The less successful agent is usually given "another chance" by being exchanged with another agent of the same type.

Essentials of Better Extension in the Balkans. In discussing the essentials of better extension in the Balkans, it is advisable to again stress the need of a comprehensive, well-balanced program which involves all phases of farm and home life for both adults and youth, in order that the development of any one phase may not be hampered by inadequate attention to related problems. In other words, increased agricultural efficiency and income should be translated into more abundant and wholesome family living.

In reorganizing the extension systems now in operation in the various countries of the Balkans, a few pivotal points suggest themselves:

1. The extension system must be established and the extension program developed with reference to existing national and local cultural programs. The Balkans are a region of great cultural diversity as exemplified by nationality, religion, and educational differences and variations. Each country has more or less important national and religious minorities, with customs and traditions which they firmly maintain. For example, in Yugoslavia in addition to national minorities there are Orthodox (Greek Catholic), Roman Catholic, and Moslem religious groups. Northeastern Yugoslavia has a natural corn-hog economy. The Moslem mountaineers import the corn as their staple food, but do not eat the surplus pork of that region or of their Orthodox neighbors in the same mountainous region. The one group herds sheep, the other pigs. Some Balkan people cook with lard, others with sunflower oil, and the people along the Mediterranean use olive oil. Male doctors cannot conduct health work with Moslem women, nor would Moslem girls study nursing or medicine under male teachers.

2. The present trend toward more democratic administration of extension should be encouraged and facilitated. When the people learn the extension system is theirs and for them, and participate in program planning and operation, success has been rapid and pronounced.

3. Local financial cooperation, however small, is essential in organizing and maintaining extension work in Balkan countries. In the past, all salaries have been paid by the national state and the national budget has provided funds for most demonstrations and improvements. The author's observations over a decade of working with Balkan extension is that the best results have been secured when part of the salaries of agriculturists, home-making workers, and health personnel has been paid from local funds. Local people in many Balkan villages have also bought grain drills, bulls, incubators, and so on, particularly where a cooperative or breeding society is available, and have contributed cash, building materials, and labor.

4. Local cooperation in studying and determining local needs should be enlisted and utilized on a democratic and continuing basis.

Allen [3] has given a moving and inspiring account of the use of local leaders in a ten-year program of uplift in rural Greek Macedonia. The book is really a handbook on Balkan extension and discusses practically all phases of a successful all-round approach "to helping the peasant help himself," a technique which has proved itself in many Balkan communities.

5. The educational nature of extension should be emphasized in setting up extension systems and in carrying on the program itself. too often the people have been forced to disinfect wheat seed, chop down infected plum trees, castrate inferior breeding sires, eliminate breeding centers of malarial mosquitoes, bring babies to a public health clinic, and so on, without being told why the program was important. One Balkan country found that two or three progressive farmers in each village would volunteer for disinfection of wheat seed and the results were so obvious that they became leaders in developing village-wide programs through the local cooperatives. Previously, legislation not accompanied by demonstration failed to produce results for the same extension staff working on another problem closely related to the first.

6. Rural youth in the Balkans requires informal methods of extension education conducted by men and women trained in the problems of working with Balkan youth. Efforts have been made in some Balkan countries to form youth organizations of the Fascist type. Even when individual clubs have been organized, results have not been satisfactory because of a lack of interesting activities. On the other hand, a number of boys can be organized in an informal course in carpentry or grafting wild fruit trees. The boys themselves usually decide to form a club for further economic or recreational activities, select their own leaders, develop a program, enlist the cooperation of their parents, and inspire boys in near-by communities to repeat the same process.

7. The extension system should cooperate with and work through local agencies such as cooperatives, credit associations, animal improvement societies, rural health cooperatives (which have rendered outstanding services in Yugoslavia), women's working groups, and religious organizations. Far too often the extension service has formed new organizations, often by compulsion through legislation and pres-

[3] H. B. Allen, *Come Over into Macedonia* (New Brunswick, N.J., 1943), pp. 239–246.

sure exerted by national, regional, and local authorities. The new organization has created little real interest or activity, and relatively effective organizations previously operating have become inactive or in many cases have ceased to exist. In other cases, the real community leaders have resented the displacement of their traditional organization, and have engaged in effective passive resistance and occasional sabotage toward the imposed organization.

8. Volunteer local leaders chosen by the people out of their ranks and given informal training by extension agents should be increasingly utilized in the extension teaching program in the Balkans. Community hatching and brooding of pure-bred and healthy chicks has been developed in over a hundred Bulgarian village communities utilizing outstanding young men given a short training course by a competent and practical Bulgarian poultry scientist, trained in American poultry science and extension.

Malarial control, building sanitary privies, and similar prospects have been carried out successfully by local men trained by a sanitary engineer. Management of breeding sires, preparation of fruits, vegetables and eggs for market, and budding trees in local nurseries have also been done by local men given a short, practical training course by an extension agent.

9. Greater and more effective utilization of extension specialists would greatly improve the Balkan extension services and provide a more balanced program. In Bulgaria, a specialist in plant protection is assigned to each of the seven provinces and maintains contact with the Plant Protection Research Institute and also the 12–15 county agents and their assistants in the province. He studies the important insect pests and plant diseases of the province and assists the agents in developing programs with the local people. An experienced practical homemaking specialist also is assigned to each province and works through local homemaking specialists while keeping in contact with the Homemaking Institute and normal school, the Institute of Hygiene, the Bulgarian Red Cross, and other agencies. In Greece, specialists from the Tobacco Institute have done outstanding work in improving the production and marketing of tobacco, which is the most important cash crop and export item of the country. A successful campaign against the Dacus fly ravaging the valuable olive crop of that country has also been conducted by specialists.

PERSONNEL. Throughout this chapter reference has been made to extension personnel. In conclusion it may be useful to make a few observations on selection and training of extension workers.

Increased numbers of men and women having a rural background or experience should be recruited and trained for extension work. Scholarships for deserving and gifted peasant boys and girls who have already displayed ability in local extension programs would be of the greatest value. City-born men and women should be given experience in rural work before becoming full-fledged extension workers. Each Balkan county requires graduates of institutions of higher learning to complete a successful year's apprenticeship before securing government employment. Practical experience in agriculture, homemaking, or health work during the training course should be supplemented by a well-planned and supervised year in a rural environment and applied especially to urban youth. Unfortunately, the city youth often does his or her apprenticeship in an urban atmosphere because this permits the apprentice to live at home and thus reduces the cost. A small subsidy for each urban apprentice would count for little in the national budget and would provide better extension workers. But the emphasis should be on rural boys and girls.

Extension workers in the Balkans must live and work under primitive and difficult conditions, hence strength and good health are essential. The conservatism and narrow moral code of the Balkan peasantry is also applied to the intelligentsia, and those who openly conflict with it will be greatly handicapped, no matter how gifted and competent they are in their field of work. Overdressed extension men and those whose social inclinations are too strong are contemptuously dismissed as "tango experts." A social butterfly working as a homemaking specialist was known as the "Parisian Mannequin" and by other less complimentary names. Parents sent their girls to her with great reluctance, fearing that she might have a bad influence on them. On the other hand, one of the most successful extension workers observed by the author wrote poetry on rural subjects in his limited spare time and spent an occasional evening dancing in places of good repute. The local newspapers printed his poetry and the schoolchildren memorized and recited it with the approbation of the elders. In fact, peasants would remark with pride that their county agent could hold his own socially with any lawyer or socialite in the city where he made his headquarters. He now holds a high position in the extension service, and, in addition to effective work in the Ministry and in the field, usually heads delegations sent abroad to conferences and is in charge of receiving visiting delegations. Neatly dressed homemaking specialists, who can make their own clothes and teach others to do likewise, have had great influence and prestige in their areas and have improved local standards, as have those who have simply but tastefully furnished a home or a room in a peasant home.

A rural public health nurse endeared herself to the local people by holding her wedding in the village rather than at her home in a near-by city, because she felt she could not leave a patient, dangerously ill. Prior to this event she had found her work difficult, as she was working with primitive and conservative people. After the wedding the local leaders became an unofficial committee to promote her health work, and the next year she received an achievement award and her region became a training and demonstration center. Her husband became a teacher in the village school and the wedding gifts from the peasants went a long way toward furnishing their home.

Preservice Training of Extension Workers. This must include courses in such subjects as economics, rural sociology, extension methods and techniques of agriculture, homemaking, and health improvement adapted to rural conditions. It will require considerable modification in the curricula and methods of teaching of the institutions of higher learning in Balkan countries.

In-service Training of Extension Workers. For many years to come a large proportion of the extension staffs of Balkan countries will consist of men and women already in service. Short training courses, conferences, and effective supervision must be provided for them. The author observed the benefits secured in several Balkan countries from short annual conferences lasting one or two weeks. The personnel was encouraged to report and lead discussions, and attended informal lectures and discussion periods conducted by competent specialists from within and outside the service. Workers must be kept informed about new developments in subject-matter fields and extension techniques and concerning special programs needing emphasis from time to time. An efficient and justly operated system of promotion and rewards for outstanding service and ability is a necessary corollary to selecting and training personnel in Balkan countries.

Extension work in the Balkans is a fundamental need. The conservatism and primitive conditions of the region make the work difficult and require well-trained, energetic, patient, and resourceful personnel who can work in cooperation with local people in helping them to help themselves. In no region, however, will local people give better cooperation and accept more wholeheartedly a sound program, suited to their needs and developed cooperatively on a democratic basis by the right type of extension man or woman.

Chapter 9 · EXTENSION WORK IN LATIN AMERICA · By Charles P. Loomis

THE GREAT DIVERSITY OF CULTURES. The term "Latin America," was invented to include all areas in the Western Hemisphere in which Spanish, Portuguese, and French are spoken. However, in the American Republics ordinarily listed as belonging to the Latin American cultural group there are dozens of different Indian languages and dialects. Also there are areas in which Oriental and non-Latin European languages are spoken. Of course, language differences should present no insurmountable difficulties for effective extension work, but the many different languages spoken by the millions of Indians and other groups in Latin America are associated with great differences in value systems, social structure, and other cultural characteristics. In addition to these cultural differences and associated with them are the racial differences. Each country differs in respect to race but the three predominant races are, of course, the white, the brown or Indian, and the black. All types and combinations of these exist in Latin America and in addition there are several hundred thousand Japanese, principally in Brazil.[1]

The total population of the Americas is estimated at 284,000,000— 147,000,000 in Anglo-America and 137,000,000 in geographic Latin America. The Latin American population is broken down by Donald Brand into 28,000,000 whites, 60,000,000 mestizos, 29,000,000 Negroids, and 20,500,000 Indians.

Since the type of Indian population in an area is closely related to the degree of heterogeneity of the culture and the level of agricultural development, no discussion of extension work and its future in Latin America would be complete without some reference to the various Indian stocks. Unfortunately, the definitions concerning who should be classified as an Indian and the estimates of the numbers of persons who are Indians vary greatly from authority to authority. The countries having relatively the largest Indian population are Guatemala

[1] The data here presented are from "The American Indian: Forgotten Man of Four Centuries," *Proceedings of the Conference on Latin America in Social and Economic Transition* (Albuquerque, N.M., 1943). T. Lynn Smith has estimated the number of Japanese in Brazil as 360,000. This is considerably larger than some other estimates. See his "Applied Anthropology in Latin America," *Applied Anthropology*, Dec., 1943; and J. F. Normano and A. Gerbi's *The Japanese in South America* (New York, 1943).

and Paraguay (where the Indian population constitutes from 65–70 per cent of the population); Ecuador and Bolivia (55–60 per cent); Peru, Honduras, El Salvador (40–50 per cent); Mexico (29 per cent); and Colombia, Venezuela, Panama, and Chile, (9–12 per cent). The lowest percentages of Indian population are found in the West Indies (Haiti, Cuba, and the Dominican Republic), Costa Rica (2 per cent), Argentina (2 per cent), and Brazil (3 per cent). Although characterizations of the many and various Indian cultures of Latin America would be pertinent to extension, the limitations of space make this impossible. Suffice it to say that the Indian cultures range from the most advanced to the most primitive, that there are Indian societies where European culture predominates and others where it has not penetrated very deeply. The diversity among the Indians is indeed great. Because of these great diversities, it is obvious that a brief chapter on extension work in Latin America can be nothing but a cursory survey or, at best, merely an introduction to the problems involved. Occasional reference is made to the Spanish American region of the United States because of its cultural similarity with many parts of Central and South America and because of the experimental extension work being done there.

CULTURAL CHARACTERISTICS SIGNIFICANT FOR EXTENSION WORK: *Great Importance of Personal Relations.* The North American agricultural technician, because he may not be interested in or familiar with phenomena outside his narrow field, is often characterized as "overspecialized" and narrow-minded. Since no one can master all fields of agricultural science, specialization is necessary for the sake of competence. In many Latin American countries where in the past the social structure has been relatively unstable, few people feel that they can become specialists. Many competent professional men, being unwilling to "put all their eggs in one basket," hold several posts in business, in government or in the occupations. Like the general farmer, they have the security which comes from knowing that if one line fails the others may not.

As is always true of unstable social situations, people devote an inordinate amount of time to personal and political relations either to retain the present position or to improve it by advancing the status of the group as a whole. Too few of the employees who are supposed to do agricultural extension work have what Veblen called the "instinct of workmanship" or the deriving of great satisfaction because a project is well done according to the standards of the profession. Too often the agent engages in "apple polishing" or does that which he thinks will give him political power rather than that which will advance agriculture. This may be related to the insecurity of tenure of those in public service and to the undeveloped state of the professions which ordinarily establish high standards of competence for their members.

Superiority and Subordination. Certain other items of social structure are peculiar to Latin American culture and are extremely important for agricultural extension work. The "peon-patron" relationship prevails to a greater extent in Latin America than in Anglo-America (meaning by this latter term those parts of the Western Hemisphere where the language and general cultural streams are predominantly English).[2] "First there is a certain submissiveness resulting in a willingness to permit, without question, both church and lay dignitaries to determine individual action. There seems to exist a sort of potential peon-patron relationship in lay affairs and in other matters of the padre, or priest, and his council are accepted with less questioning than is the case in comparable situations in Anglo-American culture."[3]

A large portion of the agricultural population of Latin America lives on huge estates controlled by a few wealthy landlords who may seldom see their plantations and know little about the welfare of the people living on them. It should be obvious that the extension work conducted among independent land-owning peasants on the one hand and the semiserfs on the other calls for different considerations. Here it is important to stress the fact that in Latin America almost every type of tenure and all varieties or patterns of superiority and subordination exist. One of the most serious conflicts between Western culture and the original Indian culture has resulted from misunderstandings and ignorance concerning Indian land tenure practices. Before successful extension or colonization programs can be developed

[2] A whole literature has grown up on the subject of differences in Latin American and Anglo-American cultures. Among the writings in Spanish of this type which have become classic, maintaining that the differences are very great, are the philosophical treatise, *Ariel* by José Enrique Rodo, and the poem *A Roosevelt* by Ruben Dario.

[3] See a discussion of this in Charles P. Loomis and Glen Grisham, "The New Mexican Experiment in Village Rehabilitation," *Applied Anthropology*, June, 1943; and Florence Kluckhohn, "Los Atarquenos, A Study of Patterns and Configurations in a New Mexico Village" (unpublished dissertation, Radcliffe College, Cambridge, Mass., 1941). Here this tendency is characterized as the "Patron Configuration."

for the Indian, the authorities must know what the Indian's system of relations to the land means to him and must realize that often ownership and landlordship in the Western sense are meaningless. An understanding of the rights, use, equities, and responsibilities to the land as related to social status is all-important for extension specialists.

Sometimes this so-called "peon-patron" relationship is so deeply imbedded in the everyday life of the people that outsiders do not comprehend it. Thus when a North American Mission Board purchased a large hacienda in Bolivia the missionaries, whose objective it was to free the serfs and Christianize the Indians, encountered difficulties when they tried to change the customary sanctions. They abolished whipping, a customary mode of punishment meted out to workers or peons by administrators of haciendas. Their Indians soon became the most lawless in the whole area near Lake Titicaca and other hacienda operators complained and threatened to intervene. To restore order, whipping had to be introduced by Christian missionaries. Of course, the missionaries were able to dispense with whipping as a sense of responsibility and competence in self-government gradually developed among the Indians, but the fact that it was required at first should indicate that there are places in Latin America where the social structure may be characterized as being of the "peon-patron" type. This is one extreme and, as stated above, at the other end of the scale there are whole areas populated by the most independent people renowned for their resistance to any worldly leader who tries to dictate to them. The people of Tepoztlán [4] near Mexico City could be used to illustrate this type. Many areas of Costa Rica, Brazil, Argentina, Chile, Colombia, and other countries are peopled by independent landowning peasants who will resist domination by feudal authority or landed aristocracy. All of these great differences in social structure are of utmost importance to those who desire to develop extension work.

Lack of Local Responsibility for Local and National Welfare. Closely related to these two characteristics is a third which makes it difficult to finance extension work in Latin America. Local land and other taxes are so low, even in the more prosperous farming regions, that usually most of the existing service institutions and facilities must derive support from the central government. Each region or locality competes to get as much federal or state assistance as possible without

[4] Robert Redfield, *Tepoztlan* (Chicago, 1930).

assessing local property so as to carry its share of the burden. Seldom do landlords with huge domains favor even reasonably high local taxes for services that would benefit the many people who furnish the labor force for their operations. It will be difficult to develop an effective extension service on the present tax structure, which does not benefit the population as a whole.

The Great Importance of Familism. Another distinction often mentioned when Latin American and Anglo-American cultures are described is the relative importance and large size of the family in the former culture. Of course, extreme familism is common and takes many forms among peasant peoples.[5] Extension and rehabilitation agencies which fail to use the large family, including as it does grandparents, children, grandchildren, and other kin, frequently overlook one of the most effective channels through which small cooperative activities may be initiated or other changes introduced into a village.

In the several typical communities of Latin culture studied by the author, more than four fifths of the families engaging in intensive mutual aid are related by blood kinship. In various Anglo-American communities studied, less than one third of the cooperating families are thus related.[6] Although there is great variation in this respect, rural familism, when measured in these terms, is stronger in Latin America than in Anglo-America. There are, of course, newly settled areas in Latin America in which familism is relatively less pronounced. Wherever it is important, unless there are counteracting influences, the nature of the functioning of committees in extension work will be different than in other areas. Familism offers great advantages in furnishing the basis for cooperation, but it is apt to be associated with the showing of favoritism toward one's relatives. This may result in nepotism, feuds, and clannishness, which make it difficult to develop professional ethics and competency among public servants.

The group and familial ties vary from the extreme group solidarity among the village people of the Ayllus of the Andes, which from the times of the Inca administrators furnish one of the most extreme forms of subordination of the individual to the group that has ever existed, to independent isolated peasant operators who dominate a few areas.

[5] For supporting references on familism, see Charles P. Loomis, "Extension Work for Latin America," *Applied Anthropology*, Dec., 1943.

[6] Charles P. Loomis, *Social Relationships and Institutions* (Dept. of Agriculture Social Research Report No. XVIII, Washington, D.C., Jan., 1940), p. 33. This gives a graphic presentation of association patterns of various cooperative groups.

After having lived on an isolated holding in an area of independent small holders, one appreciates the multifaceted control of the familistic group over the individual in an isolated Ayllu village. Among the former, a sense of independence often develops in which the individual resists becoming "beholden" to anyone. In the Ayllus, where most property was once owned in common by families and where most decisions are made by the group or its representatives, individual independence is minimized.

The Importance of the Village. Most visitors to Latin America are impressed by the sameness of the forms of the villages, towns, and cities and the prevalence of village communities in rural areas. This sameness of the urban scene is emphasized by James: "In most of the Spanish colonial cities the dimensions of the blocks, the width of the streets, and even the arrangement of the government buildings and the plaza were all standardized . . . features which characterize Spanish cities from California to the Strait of Magellan." [7] The rural investigators and visitors in both Spanish- and Portuguese-speaking rural areas are also impressed with the extent to which the standard form prevails. However, as Smith [8] has shown for Brazil and as Taylor [9] has shown for Argentina, much of the Latin American rural home life goes on outside nucleated or line villages.

Not in all cases in Latin America have Spanish and Portuguese village patterns completely displaced the original precolonial form of settlement. In many areas the original Indian settlement pattern remains. Despite important differences between Latin and Indian cultural items such as land tenure, family structure, and the like, it is sometimes remarkable how little change has been necessary to integrate the two patterns. Most of the precolonial Indian communities were village communities. Thus, since Spanish and Portuguese farmers came from cultural areas where the village form of rural settlement prevailed, it was only natural that with the exception of some Brazilian areas and some temperate zone regions which in precolonial times

[7] Preston James, *Latin America* (New York, 1942), p. 182.

[8] T. Lynn Smith, "The Locality Group Structure of Brazil," *American Sociological Review*, IX, No. 1 (Feb., 1944). Here Smith states that Brazil, which constitutes about one half of South America's populations is not really a farm society. Villages are most frequently composed of people who do not till the soil. Workmen's families on large holdings are grouped.

[9] Carl C. Taylor, "Rural Locality Groups in Argentina," *American Sociological Review*, IX, No. 2 (April, 1944). Argentina, according to this description, has all types of settlements but the village peasant farm operator is rare.

were occupied by nomadic Indian populations, the village type of settlement prevails today. The most common form of rural village or town consists of rectangularly arranged streets and a plaza, in which the church and market are found and about which stores and official buildings are located. It is obvious that an extension agent who works among the farm people of one of these villages should orient his program to the village pattern.

As previously indicated, throughout Latin America, familism is the leitmotiv of village life. Family politics run the social institutions. The businessman with large family connections is more apt to succeed, other things being equal, than one with few relatives. Whole villages or village factions are often family groups. Hence the importance of the previous discussion of familism.

The present author assisted in the analysis of the methods used by the United States Department of Agriculture in family rehabilitation in "El Pueblo," a Spanish-speaking village in New Mexico, in the hope that the village approach would have wide application in the countries to the south. Through intensive home and farm management supervision, remarkable improvement in the level of living was attained in a short time and it was concluded that the village approach in which the interrelations of families was the prime focus was better adapted to rehabilitation and extension work among the Spanish-speaking villagers of New Mexico than was the individual family approach. The village structure was used to lengthen the hand of the individual supervisor. The ditch associations, the small family-friendship cooperatives, the church and all the other organizations of the community were used to implement the rehabilitation and extension program. To use an analogy, the supervisors used the "handles" in the villages which would multiply the results of their efforts. When "handles" were not available they were created. In this sense the Community Council and the Livestock Association were handles which the supervisors helped forge to increase the effectiveness of their program.

The Church. By far the most important formal social organization larger than the family in most Spanish-speaking villages is the church. The importance of this agency and the importance of eliciting the cooperation of the priesthood in extension and rehabilitation work cannot be overemphasized. In areas where the Catholic Church is important, a priest who is interested in improving the technical agricultural practices of the people can accomplish in a few months what it would normally take others years to accomplish. Of course, many priests are not interested in agricultural extension work and many areas in Latin America do not have priests. Nevertheless, rural Latin America is overwhelmingly Catholic and the fine work of a few priests in the organization of cooperatives and other activities of an extension nature demonstrates what an enlightened priesthood can do in Latin America.

Attitudes toward Money. With considerable evidence it has been claimed that most rural Latin Americans (and many other people who were relatively isolated during the long period of the development of what is called modern capitalism, or modern commercial and industrial enterprise) have attitudes toward the use of money which differ from those of Anglo-American farmers. While rural Anglo-Americans were disposing of the various vestiges of feudal and communal land tenure systems the Spanish Americans and many Indian groups held things in common or under feudal tenure. Because they have lived for centuries outside the realm of an industrial and highly competitive money economy they cannot be expected to manifest the modern businessman's attitudes toward money and the various aspects of money, such as interest. When money is not a value in and of itself the extension agent may not want to argue that the change he advocates will pay. Other means of motivation may be more important. Even though the general outline of this characterization may be supported, there are great variations in this respect among Latin Americans. One must not assume that this attitude toward money is universal in Latin America.

Changes in Agriculture Spread Slowly. The effectiveness with which extension work may be carried on is closely related to the phenomenon which cultural anthropologists call diffusion. A new invention or an improved practice, such as hybrid seed corn and its utilization, may spread very rapidly in one cultural or geographical area; in another area, it may spread more slowly. The rate depends, in large measure, upon the integration of the culture. In Latin America, both cultural and geographical factors restrict rapid spread of improved practices. Populations are frequently heterogeneous and the people of integrated cultural areas are conservative, resist change, and are suspicious of new traits.

Even if this were not the case, there are important physical factors which retard change. In the mountainous regions human habitation is clustered in pockets and valleys at extreme altitudes segregated from other nuclei by arduous passages or impassable mountain barriers. In the heavily-treed regions of the *selva,* sparsity of population and rankness of jungle growth are two factors which slow up cultural diffusion. Since most of the mountain and jungle peoples live within an area included in a band 10 degrees on either side of the equator, it is interesting to speculate on the importance of geographical factors as related to the retardation in the field of agriculture. Certainly in this area live people who are among Latin America's most primitive agriculturists. This is true even though the geographical distribution of the most advanced precolonial cultures was almost the reverse of what it is today, in that the great civilizations occupied the plateaus of the central area between the northern and southern temperate zones then occupied by the least civilized peoples. In reality modern agricultural methods have, with some few exceptions, scarcely penetrated the jungles of the tropics and the heights of the Andes.

As the future spread of modern agricultural methods is contemplated and its relation to extension anticipated it is important to consider several types of cultural diffusion which have taken or are taking place. Mexico's student of Latin American literature, Francisco Monterde [10] has noted that the trends and influences of various schools in literature read by the intelligentsia followed definite patterns in their spread. For instance, romanticism and modernism in poetry began as succeeding waves at the northern and southern extremes of the cultural area at about the same time and gradually spread with a lag of thirty to fifty years toward the equator or the countries inhabited by large Indian populations living in the jungles or at high altitudes.

If agricultural developments followed the same pattern, Argentina, Chile, and Uruguay to the south, and Mexico and the Caribbean area to the north would adopt or invent new practices first, after which they would spread toward the equatorial jungles and plateaus. Mexico's anthropologist and sociologist, Manuel Gamio, believes that the earliness with which social revolution came to Mexico is due in large part to Mexico's proximity to the United States. Mexican laborers

---

[10] Data concerning these observations, which are to be published in a forthcoming book, were presented to a class given in the National University of Mexico in 1942 attended by the present author.

who had worked in the States, even though not always gaining the highest wages, found it difficult to return to Mexican conditions of labor. It is interesting to note that other cultural traits spreading over Latin America also arrived last in the tropical and semitropical areas, where the various cooperative experiment stations and extension services of the United States and other American republics are located.[11] In an area with so many complicated cultural, biological, and physical factors it is impossible to predict with exactitude the course of important events but the above considerations suggest possible trends.

Agricultural Extension and Research Are Needed. Every country in Latin America could raise the level of living of its farm people through disseminating improved practices and planting and breeding stock which are now available. Although there are areas in Latin America where large estates have adopted the most modern equipment, stock, and methods, the majority of the people who work the soil use little farm equipment more complicated or effective than the hoe, foot plow, or machete. Little research is directed at solving the poor man's problems. For instance, little effort is being made by agricultural scientists to develop a more efficient stirring plow to be drawn by oxen or to improve guinea pigs, a source of meat for the poorest Indians of South America. Both research and extension for the little fellow are needed and it is important that in many areas the little man, though ignorant, wants this type of assistance. The observation of T. Lynn Smith for Colombia and El Salvador agrees with that of the author in respect to the willingness of the peasants to use and support extension services.[12]

Generalizations Are Difficult. In fact, the above discussion will have accomplished its objective if it has shown how difficult it is to generalize meaningfully about so many diverse countries and cultures. There are certain general tendencies, but no absolute differences. There are few cultural and mental traits present in one area which are not present in some degree at least in the others.

[11] For a description of this cooperative work in experimentation and extension see Ross E. Moore, "What Shall the Americas Grow?" *Agriculture in the Americas,* May, 1943, and Charles P. Loomis, "Developing a Permanent and Stable Supply of Needed Agricultural Materials," *Applied Anthropology,* Sept., 1943.

[12] Charles P. Loomis, "Extension for Tingo Maria, Peru," *Applied Anthropology,* Dec., 1944 and "Agriculture in the Americas," *ibid.,* Feb., 1944.

TYPES OF EXTENSION IN LATIN AMERICA.[13] It is almost as difficult to generalize about activities which may be classified as extension work in Latin America as it is to generalize about social structure and attitudes. Thus in the Spanish-speaking areas of Southwestern United States extension work such as that described in the chapter by Dr. C. B. Smith is carried on by both the Extension Service and the Farm Security Administration.[14] In Puerto Rico the extension and rehabilitation patterns of the United States were introduced and are gradually being fitted to the local needs. In some countries, particularly Cuba and Venezuela, some aspects of the United States Extension Service have been introduced. In other areas, particularly in the poorer and smaller countries of Central and South America, scarcely anything which might be called agricultural extension work is to be found. The work in Latin America may thus be said to range from that of areas having most complete institutionalized systems to areas completely devoid of extension work. On the other hand, we may say that practically every type of activity which in this book is defined as extension work is found somewhere in Latin America. Of course, outside of the possessions of the United States, no single country has an organization comparable to our Extension Service; the word extension, or a literal Spanish translation of it, is seldom used in the sense that it is used in the Continental United States and Puerto Rico. Such activities as exist are found in different ministries and departments of the governments and some are carried on by nongovernmental organizations.

Veterinarians and Regional Agriculturists. One of the most common types of extension activity in Latin America is that of the veterinarians and regional agriculturists or agronomists. Most frequently these men are employed by a national ministry or department of agriculture and serve a prescribed area, but sometimes they work for states or provinces. The agronomists are in some instances attached to local experiment stations, but in others they are not. Few veterinarians or agronomists spend the major portion of their time in demonstration work or instructing farmers individually or in groups. Generally, an

[13] For this section the author is indebted to Ralph Allee, T. Lynn Smith, Carl C. Taylor, Nathan Whetten, Philip Green, and the various experts in the Regional Branch of the Office of Foreign Agricultural Relations who prepared statements on individual countries, but takes full responsibility for the statements here presented.
[14] Charles P. Loomis and Glen Grisham, "The New Mexican Experiment in Village Rehabilitation," *Applied Anthropology,* June, 1943.

overwhelming amount of the attention of these and other men who could do extension work is consumed by regulatory and police duties. However, with proper training, facilities and larger staffs, the services to which these agriculturists are attached could in many cases develop effective extension programs.

Specialists and Campaigns. Specialists who put on campaigns are also fairly numerous. In Argentina the specialists of the cotton board are reported to be among the most effective agents. Brazil and Colombia also conduct extension work on the campaign basis. These campaigns usually lack community orientation. Often a series of specialists bombard the villages, each attempting to "sell his wares" without enough consideration for the over-all needs of the people.

Extension and the Schools. Extension work is also carried on through the school systems of many of the countries. This is particularly true of Mexico, Chile, Colombia, Argentina, and Peru. In some countries, clubs composed of students carry on projects and receive practical instruction from teachers and other governmental officials. The Ministry of Education of many of the countries has its own vocational agricultural programs carried on in both the normal schools where agricultural teachers are trained and in the lower-grade schools. Schools and ministries of education also organize fairs and exhibitions in a number of countries. In some of the countries the agricultural schools carry on a type of extension work, such as that of the Brazilian schools which have Farmers' Weeks where agricultural methods, planting stocks, livestock, and machinery are demonstrated or exhibited.

The Mexican Educational Missions. In Mexico, the Federal Department of Education has 25 Rural Education Missions which are carrying on extension in isolated regions of the Republic. The missions centrally located headquarters, attempt to extend their program to a surrounding area of about a dozen villages. Approximately one to three years are spent in each area after which the mission moves to another center. Each mission includes a director, with teacher training and agricultural experience, an agriculturist, who is a graduate of an agricultural college, a nurse, a social worker, a recreation specialist, a musician, a carpenter and, two or more teachers of trades or crafts. The missions attempt to orient their programs to the needs of the people through the advice and discussion of a Committee of Economic and Social Action. This committee is made up of local residents and to it is entrusted the coordination of the total improvement of the community. This program has the advantage of being oriented to the needs of the local people, being equipped to cover a wide range of needs and being organized so that its influences may spread over a large area beginning with villages which are most receptive. However, the missions do not remain in an area long enough and fail to follow up their program.

Extension and Private Agencies. Extension work in Latin America is also sponsored by such nongovernmental agencies as the National Federation of Coffee Growers, the Stockmen's Association, and the National Agricultural Society in Colombia, and the Federation of Coffee Growers and a semiprivate agricultural institute in El Salvador. The South American Fruit Company, various railroad companies, the Agronomic Society of Chile, and the United Fruit Company in various of the Central and South American countries carry on an instruction program in agriculture. Also in some countries such as Mexico, Paraguay, and El Salvador, banks have developed both experiment station and extension programs.

Demonstration Farms and Ranches. Demonstration farms or *granjas* are fairly common in Latin America although, as is the case in many other areas, they often have little more influence than an exhibit of the artifacts of the cave men or pictures of the world of the future as seen through the eyes of an H. G. Wells. Frequently in areas where the people use the foot plow, the "demonstration" *granjas* are equipped with tractors pulling combines and huge many-bottom plows, and try to keep accurate farm management records. To expect such "demonstrations" to change the people is like expecting Mark Twain's jumping frog to hop over the moon.

The Church Missions. Probably the most important type of nongovernmental extension work carried on among the Indians in Latin America is that of the various church missions. Notable examples of this work is that developed in the Sierra among the Indians of the Andean countries by the Salesian Brothers, the Seventh Day Adventists, Canadian Baptists, and the Evangelical Union of South America. Also various Catholic Orders other than the Salesians have been doing extension work as a part of their mission programs for many years among the Indians.

The extension work of various church missions of the Andean countries and the local rural schools, particularly those in Mexico,

Colombia and Chile, and the work of the Y.M.C.A. in the area of Tepoztlán, near Mexico City, tend to be centered in the local communities and to be oriented to the local needs of the people. This stands in sharp contrast with the work carried on by most of the other agencies. In fact, one of the greatest weaknesses of the work carried on by the regional agronomists and veterinarians is that their duties are all too frequently spelled out in detail on the national level, making it difficult to meet specific local needs. Another shortcoming of the extension work of governmental agencies is the tendency for various specialized bureaus with overlapping or competing programs to push their work with little coordination either on the national or local level.

Extension and Colonization. Where colonization work is going on, as in Argentina and some of the Andean countries, there seems to be more orientation to local needs, which leads to a more integrated program than is found in areas less recently settled. This is partially due to the fact that there is official recognition that the development of the colonies cannot be blueprinted because too little is known about farming methods in the new areas.

Home Economics in Latin America. Few areas of the world could gain more comfort and health from the introduction of improved home practices. However, little work is done by extension specialists in improving the home and improving the efficiency of the housewife except in Puerto Rico and on some haciendas in Chile.[15] Since few women enter the professions, this field cannot be expected to expand rapidly but the generalizations made below apply to it as well as to that of improving farming methods. The 4-H clubs, are, however, spreading to other countries.

LOOKING AHEAD TO THE DEVELOPMENT OF EFFECTIVE EXTENSION WORK: *General Adjustments Which Are Needed.* In most of the Latin American countries, the rigid class structure complicates the development of extension work and is reflected in the various officials who carry on the work for the government. The graduates of the agricultural colleges are usually sons of well-to-do land owners or farm managers who have never actually done farm work. These graduates usually have the disdain for physical labor characteristic of the

[15] Elizabeth S. Enochs, "Transformations in Remote Places," *Land Policy Review,* VI, No. 2 (Summer, 1943).

wealthier classes. This general disrespect on the part of many extension workers for, and their lack of experience with, actual physical labor makes it difficult for them to organize effective demonstration work with the small farmers who must actually work with their hands. On the other hand, the large estates themselves can afford to hire well-trained agriculturalists. Thus, it is obvious that the families of the small farmer and the peon on the large estates are in most need of the assistance of extension workers in the matter of food production and preparation. One of the greatest needs in Latin America is the development among agricultural workers of a professional attitude comparable to that of the British Agricultural Organizer, the American County Agent, and Farm Security Supervisor. This would promote the development of a relationship of mutual respect and helpfulness between farmer and trained technician. Each would learn from the other; respect for the agent's ability in his field of competence would partially overcome the class barriers.

The general structure of most of the Latin American countries does not permit much voluntary local action in determining national policies or in electing national officials who function in local communities. This same principle carries over into most of the extension work done by governmental agencies. The bureaus are not so well-established as to permit the study of the farmer's needs and the development of research and extension programs to meet actual needs. As in the case of many centralized governments, no effective mechanism exists through which the wants and desires of the people are brought to bear upon the general functioning of the bureaus. Therefore, considerable change in organization is needed. Extension work must be recognized as a two-way affair. The needs and experiences of the farmer must be carried to the experiment stations and colleges just as the knowledge accumulated in these agencies must be carried to the farmer.

More local responsibility would lessen the tendency of agricultural specialists and bureaus to use local farms and communities as fields within which to develop their own specialized interests, irrespective of the local needs. There are many advantages inherent in that type of extension system which makes the agent responsible to various local units with some control over his work. Of course, this specification must always be coupled with another: that the local agent, because of his professional attitude, will never exploit his position to gain political office or power. As the agent develops this attitude he will render competent professional service directed toward the felt needs of the people.

As one want is met, others may be expected to develop around which new programs must be built. This general approach differs greatly from the health and sanitation brigades of the Andean countries. All too often these "flash" programs, even when supported by movies and other devices, do not carry through and in the long run accomplish relatively little. Even the Mexican cultural missions would no doubt be strengthened by a strong follow-up program. It is recommended that extension work be integrated into the everyday life and culture of the local community in such a manner as to get the maximum effect for each project, which must be followed up and carried through.

Extension work in Latin America, being in the formative stage, could well avoid some mistakes made in other countries. In various parts of the world, experiments have demonstrated that most progress is made when the different agencies work together on the whole range of needs and wants of the local community. If different, poorly-integrated administrative units are established for all activities of separate bureaus much competition, overlapping, duplication, and frustration of work and purpose and general inefficiency are the inevitable results. It, therefore, is common sense to recommend that the extension program should be established in such a manner that the extension agent utilizes the resources of all bureaus to solve rural problems. He should be a well-trained agriculturist but he should not be so highly specialized that he lacks sympathy with and understanding of the health, recreational, and other essential needs of community life.

In Latin America, as well as in other areas, there is often a tendency for lawmakers and bureau officials to attempt to spell out all that should be done in the future by a given agency which is being established. It is highly important that rules and regulations and bureaucratic red tape should not be permitted to distort or hamstring the normal development of extension work; which, because it should be oriented to local needs, cannot be blueprinted in detail on the national level either in formal legislation or bureaucratic regulations until experience in the community has developed a desirable local pattern.

All the agencies which are carrying on extension work, whether governmental, church, or private, could improve their programs if certain principles were followed. These principles will be discussed one after another and their relevancy to Latin America indicated.

1. Work should be started in communities where entry can be made comparatively easily. The effectiveness of this principle has been demonstrated by Spencer Hatch in India and Mexico [16] and at the present time the International Missionary Council is studying the Andean countries to determine where to enter with an agricultural mission program. The principle may appear obvious but it has often been ignored, especially by agricultural missionaries. It has been compared to infiltration in military tactics. To attack where the greatest resistance will be offered is often foolish because the communities offering this resistance may be easily won over later when they see how effective the methods are in the communities where an entry can be gained easily.

2. Gain a thorough knowledge of the main values or pillars of the local culture before launching any program of action. This means that rural extension services, rural schools, and agricultural missions which carry on extension work can learn much from the cultural anthropologists and rural sociologists who have made studies in Latin America. If the values (including beliefs, attitudes, taboos and the like) are not known, mistakes which may delay programs many years are apt to be made. In one area in Latin America a government agent from the United States almost caused a mutiny among his workers by ordering that mango trees, which through the ages had been respected producers of a vital food, be chopped down along with other trees to make room for a crop needed for the war. After considerable time was wasted trying to get the balking natives to work, the agent realized that cutting down mango trees violated the mores. Thereafter the mango trees were permitted to remain standing.

3. For demonstration purposes choose a site that is advantageously located. Roads, watercourses, and all types of channels of communication are determining factors in the choice of sites for demonstration projects. Isolated settlements are common in the equatorial jungle and mountains. Other things being equal, the communities in the larger nuclei of population should be entered first in order to get the maximum spread. Not far from Tepoztlán, Mexico, Hatch chose a site for his housing and farm demonstration project. Through the narrow valley in which the project is located the people of the sur-

[16] Spencer Hatch, *Up from Poverty in Rural India* and *Further Upward in Rural India.* Also see *Applied Anthropology,* June, 1943, and Sept., 1944; also: "Reconstruction in Mexico," *Agriculture in the Americas,* March, 1944.

rounding villages must pass to get to the trade center in Tepoztlán. In only a few years, they have stopped to watch, question and adopt many items from the demonstration center.

4. *The needs of the whole community should be met.* No one faction or group should receive all the benefits of the program. Thus at the resettlement colony of Tingo Maria, Peru, more attention was given to the mestizo farmers than to the poorer Indians. Already, after only a few years in the colony, quite serious diseases have developed among the Indian settlers and these diseases have become a menace to the whole colony. If extension is to be effective in Latin America the landlords of the great plantations and large estates must realize that it is to their advantage to improve the health and well-being of the poorer laborers and croppers. All too frequently a small group of experts who have specialized in a single crop put on a campaign to get the people to change practices involving this crop or to adopt it on a broad scale. Technical professional groups are apt to be like evangelical religious cults in their ardor to advance their specialty. They push it often disregarding the other needs of the people.

5. *Demonstrate the need and practicality of the new program before trying to push it.* In the rehabilitation program in the Spanish American community in New Mexico studied by the author, the water in the ditches and river from which the people procured their supply for drinking and home use was found to be contaminated. Villagers were required to dig through fifteen feet of solid rock to get well water. Later the wells were reported by the State Health Service to be contaminated. The health authorities then chlorinated the water and, as a result, the villagers began again to get their water from the contaminated ditches and the river. They are still doing this because they have not developed a felt need for pure water.

6. *Bring together in a familiar environment people who already know one another.* Some of the schools in Latin America do violence to this principle. They take the Indian children out of their communities to live in a new and strange environment and do not permit them to return home for a period of years. Many never return to the villages and the few who do return after such education find it difficult to apply what they have been taught. Instead, they are forced by their conservative elders to conform once more to the life of the village, where social pressure is very great against anyone who seeks to be different. The several church denominations which work directly

in the villages may not change the children so rapidly, but the changes are more lasting than in the interne type of training which divorces the child from his own people. When leaders in the community try out new practices and these prove to be successful others will follow. This is not always true when strangers adopt or bring in new practices.

7. *Start with projects that are important to the farmer and whose importance will be easily demonstrated to him.* If it is desired to change a practice which results in "bucking the mores" much time will be required. The extension or other service can attain status for itself more rapidly by working on changes which demonstrate their value without arousing emotions. Missionaries who begin by attempting to change morals before they have demonstrated the value of the new practices they wish to introduce may postpone the attainment of their objective. They should begin with something to which the people are more receptive. Protestant missionaries have been successful in getting thousands of Indians to give up chewing coca leaves (from which cocaine is made) by substituting American soft drinks. However, coca has been so closely tied to the social values and customs of the people since Inca times and is such a strong habit-forming drug that the change could be accomplished only after the missionaries had carried on an educational campaign and established themselves in the area through their medical educational services.

8. *Start with what the people have.* Many of the experiment stations located in the poorest areas in Latin America work with modern mechanical equipment, whereas the local inhabitants use hand tools and are too poor to buy mechanical equipment. Some church stations demonstrate agricultural methods to their internes with equipment completely beyond the farmer's reach. Purebreed livestock is exhibited and local stock disparaged when an improvement breeding program based on the hardy local animal is what is really needed. Agencies content to start with the people where they are, gradually leading them to the use of better practices, are likely to make the most progress in the long run.

9. *Let the program evolve from the people and remain their program.* Programs brought in from outside by polished-booted agents, who would not stoop to soil their hands, may be looked upon by the common people as theoretical or as "book" agriculture. An agent stands a much better chance of having scientific practices accepted when he establishes himself with the farmers and works with them,

and takes their own problems to the experiment station for solution. When a disease or a pest attacks crops or livestock, the farmer or committee requesting a cure or control will probably put it into operation immediately. No salesmanship is required for what the people themselves want and ask for. Many an agent has delayed the acceptance of a practice or implement by pushing its adoption so hard at first that conservative leaders who did not understand its advantages took a public stand against it. When this happens, the agent must design a face-saving device before adoption is accomplished.

10. Utilize local leadership appropriate to the given situation. No effective extension agent overlooks the necessity of extending his activities through the use of leaders who keep him in touch with the local problems and felt needs. They may form his advisory committee, channel problems from the whole area to him, and carry back the solutions and improved practices.

When a new implement, plant, or animal is introduced it is often advisable to make it a symbol of prestige by giving it to a leader. Sometimes this may be accompanied by the customary ceremonies befitting such transmittals. Of course, precautions to prevent one group from monopolizing the item must be taken. On one of the Indian reservations in the United States a vitamin deficiency was overcome by Indian Service authorities through the medicine men who administered pressed, green pine-cone extract through a ritual. If the preparation had been more palatable and the people could have later been taught to prepare their own medicine, the adoption of the new item would have been more effective.

In Peru a Protestant physician established a small but well-equipped medical and surgical clinic far up in the Andes, many miles from any hospital or adequate medical facilities. Local doctors prevented the surgeon from getting his license to practice. However, when they attempted to have him removed, the local people, including their leaders, flooded the President of the Republic with letters. The President himself visited the clinic, found that the local people were getting excellent service and the license was forthcoming. Local leadership combined with the outside leadership overcame opposition. One of the most interesting projects ever designed to test this general principle is the Indian Lay Health Program for Nicaragua and Ecuador of the Inter-American Indian Institute. Arrangements have been made to introduce simple medical practices and medicine to the Indians

through their medicine men and other local leaders. Health and sanitation surveys are made to diagnose the type of aid which is needed
and can be introduced by this method, after which the local people
are given the required type of training to carry on the program. The
program is, to date, experimental but the response has been very good.

To fail to use these structures in extension work is akin to the act
of a logger throwing away his cant hook and rolling logs with his
hands, or to that of an auto mechanic kicking aside his jack in order
to lift the front or rear of a car by "main strength and awkwardness."
The existing cooperative structures and the pertinent leaders are
"handles" for the good extension agent. When the appropriate cooperative structure does not exist he may have to organize one out of
other structures, but he should build on existing relationships.

Chapter 10 · EURO-AMERICAN RURAL SOCIETY · By Carle C. Zimmerman

Euro-American [1] rural life has important characteristics which differentiate it from folk and preliterate society and, from the extension point of view, it also has significant internal differences. One such difference is that which exists between European and American rural life generally. Others are differences between northern and southern Europe, between different political regimes, between the Northern and Southern states in America, between the white and colored races, and the large farmer and the less well-to-do country people. Other and minor variations in regional culture cannot be considered in this general introductory sketch.

General Traits of Euro-American Rural Culture. The following general characteristics separate Euro-American society from folk or preliterate cultures.

1. Euro-American society produces goods primarily for market or for sale. While local consumption of goods produced at home is found everywhere in agriculture and rural life, market production is much more a characteristic of Western society than of others. Associated with market production, and having a decided relation to it, is the relatively easy access of the city to the country and the country to the city in Euro-American society. The Euro-American rural dweller, because of roads, railroads, automobiles, mail service, and other means of communication, has contact with the city more readily than the folk peoples of Asia and other regions.

Further, the Euro-American rural dweller responds to the city and its values much more rapidly and more rationally than does the "folk" person, who produces what he likes and sells his surplus to the city person needing or desiring it. The Euro-American farmer tends to know what the city wants and tries to produce to meet these desires. If the Boston market wishes brown-shelled eggs, these are produced.

[1] Euro-American as used in this chapter is more or less coterminus with Western Society, or the commercial, trading and industrial civilization which has developed in and spread from European Civilization in the modern age. Australia, New Zealand, Canada, most of South America, and several other regions not discussed in this analysis belong to it. The lands with Spanish and Portuguese cultural backgrounds and the Anzac regions require particular study on the spot.

On the other hand, if the New York market wishes white-shelled eggs, those selling in that market will provide that type. In England (before the present war) the farmer knew that the city person wanted whole milk of a relatively low butterfat content for his tea. That he made available. On the other hand, the Continental peasant, at least in northern Europe, knew that the city wanted potatoes and fats, and was governed accordingly. In southern Europe, the country person knew that the city wanted hard wheat for bread or macaroni. This he tried to produce. Many country producers in Italy have always fed themselves primarily on cereals made from maize, but they specialize continually in producing wheat for sale to the city. The folk person produces what *he* wants; the Euro-American rural person tends to produce primarily what the city wants.

2. *Rural life becomes standardized in terms of the city.* Another related trait is the common use of the same units of measurement and of value by the city person and the country dweller. This is reflected in market prices, universal systems of measurement, common standards of desire and of living conditions and in similar cultural milieu. The folk person uses the system of measurement of his village. The village basketmaker is an artist more concerned with the beauty of his weaving and the wearing qualities of the product than with any standard specifications as to size. As a result, the weights and measures of the "folk" sometimes vary from village to village and from person to person within the village. The folk-peasant trades by direct observation of values given and received. He is a "swapper."

But in Euro-American rural life universal systems of weights and measurements have been accepted by the country people so that products are sold by the pound, the foot, the gallon, or the kilogram, the meter, and the liter. Land is sold by the acre or hectare. These "universal" systems of measurement give a meaning to market prices which infiltrates to the country dweller. Statements of grain prices in terms of bushels or metric tons have a meaning to the Euro-American country person because he evaluates his product in those prices and quantities. Among the folk and preliterate peoples, such common standards have little meaning.

This universal system of measurement between the city and country people reflects itself through all the life conditions. When the farmer hears that someone produces a bale of cotton to the acre as compared with his one-half bale he can be made to understand what that statement means. Out of that arises the fact that cause and effect have a deeper meaning to the Euro-American rural dweller; he is learning to think in abstract terms.

An illustration of this is the north European farmer producing dairy products. He can understand an extension worker who points out that a cow grazing in the field consumes in the process just so much energy that could be turned into milk if the cow were stall-fed. He can realize the difference between the upkeep of the cow and the output of milk. He comes to view the cow more as a milk machine and will, under some highly rationalized situations, feed the cow to a milk-production level where the losses from tuberculosis would offset any gain in milk production from more intensive feeding.

A good illustration of the same characteristic capacity to understand cause and effect and to rationalize agriculture is the poultry business in America. In this industry in recent years practically complete scientific production has spread itself not only among the large producers but among many smaller ones. Large hatcheries produce the chicks, all from single specialized strains (white egg layers, brown egg layers, market fryers, and so on) and sell these chicks through the mails to those who raise them. They are then fed not merely "grain" but specified amounts of proteins, carbohydrates, vitamins, and minerals. The rural dweller and agrarian producer becomes a manufacturer dealing as rationally as possible in the products of living stuff.

This system of values spreads to the standards of living and the measurable formal standards of life. The rural dweller wants his house painted, wants factory furniture, running water, bathing facilities, evenly heated homes, academic schooling, books, reading matter, and all the accepted values of the city. This means that, to a considerable extent, these common standards of desire can be used to stimulate changes in habits of the Euro-American country person. Out of this arises the value system known as "parity." Parity, or the feeling that the country man should have all that the city person has, is an American word, but its meaning under other terms has spread throughout European agriculture and country life.[2] In England it was reflected

[2] See Charles and Mary Beard, *Basic History of the United States* (New York, 1944), Chapter XXVII; Carl T. Schmidt, *American Farmers and the World Crisis* (New York, 1941), Chapters III and IV; John B. Holt, *German Agricultural Policy 1918-1934* (Chapel Hill, N.C., 1936), Part IV; C. A. Anderson, "Sociological Elements in Economic Restrictionalism," *American Sociological Review*, IX (Aug., 1944), 345 ff.; V. L. Parrington, *Main Currents of American Thought* (New York, 1927), Vol. III.

before the War in protectionism for agrarian prices and in the application of all national acts regarding minimum wages and working conditions to agricultural and village labor as well as to industrial and city labor. On the Continent it has been reflected in the rural application of the same types of legislation, whether affecting prices, wages, school systems, rural electrification or other matters. Consequently the appeal of "parity," or making the country like the city and the countryman like the city dweller, plays a much greater role as a general concept to motivate changes in behavior among these people than it does among folk.

The folk person feels that this is his costume, his basket, his native habit or form of behavior. To these he is attached, and without them he feels strange and unclothed, either physically or mentally. The Euro-American is more apt to feel that he has "rights" to be like other persons in the greater cultural milieu and not confined to village habits and ways. The mountaineer wants to give up his native costume in Europe. The costume is all right for village day or for ceremonials or for the waiter at the "rural" beer garden in the city, but it is not for him. If the special hill troops in Italy want to wear the mountain costume, that is all right, but the country dweller wants to dress as the city person. He has moved from village perspective to the "right" to be part of the nation; from being a quaint, picturesque character to being a citizen.

3. The Euro-American farmer views his land impersonally. A next important difference between the folk and the Euro-American rural dweller concerns the land and the attachment between person and land. To the folk, the land is sacred. It is home. It is a strange something which nourishes him, protects him from uncertainty and gives him a place in life. In many places it contains good and bad spirits which must be praised or propitiated. The connection between family, person, land, and home are viewed as everlasting.

This is not true for the Euro-American rural dweller. To him land is a residence; something with pecuniary value to be ascertained roughly by a formula of expected economic production. Improvement of the soil or the preservation of its fertility is something he weighs in terms of economic return. If he is a share cropper in the Southern United States, he is unwilling to do much for the soil because he will probably be moving next year. He feels that if he improves his place, the landlord will try to get rid of him to share the later benefits. Instead of manure, which might give slow returns in soil fertility this year but lasts for several years, he prefers fertilizer which tends to yield more of its results in crops immediately.

The folk person, on the other hand, will pick up the rocks about his fields because it pleases him to do so or because he can use them to build a little monument to worship the spirits he thinks are in his soil. The farmer wants to know if the economic results will pay him wages for his work, as illustrated by the Texas farmer who will not rid his pastures of cacti and other rank growths unless labor is so cheap that the gain exceeds the expense.

But being impersonal about his lands and his home place, the Euro-American moves more rapidly from place to place and less often leaves behind him a line of children to carry on the name. The local and village milieu is continually broken up. This enhances opportunities for the infiltration of urban and general systems of economic value into the country as opposed to the local tradition and the magic found in the folk village. The Euro-American has a residence instead of a home; he has a place instead of his land; he is an individual rather than merely a member of the family or clan.[3] Hence arise the attitudes which lead the Euro-American to dissipate the resources of his land if it does not pay him well to conserve them, and that within a relatively short period. Witness the spoliation of the soil in many of the American states.[4] Witness the decline of English agriculture from the late Middle Ages. The same spirit still exists, but less potently in Europe (including England) than as in America.

4. *The Euro-American farmer can act independently of the clan or village group.* As an individual the Euro-American can act for himself. He can buy land, sell land, improve it or not improve it as he wishes, and change his crops or seed, by consulting only himself. Land titles are measured and registered. Land sales are of much more common occurrence. There is no great weight of tradition tending to keep the Euro-American in the "groove." This has advantages and disadvantages from the extension point of view. A given farmer can be taught an improved practice and he can adopt it himself. Then his more reluctant neighbors can actually see the results. If he grows a better type of seed, they can get some from his surplus in a few years.

[3] See H. F. K. Gunther, *Das Bauerntum als Lebens—und Gemeinschaftsform,* Chapters II and III.
[4] Charles and Mary Beard, *Basic History of the United States,* p. 415.

If he has a better bull, they can buy calves from him. If he uses new machinery, they can buy it, too, borrow his when he does not need it.

The Euro-American can work more as an individual without consulting all his relatives or holding a clan meeting. This is aided and abetted by the fact that the lands are held as individual property and not as an undivided unity. This can best be explained by an illustration from the rice sections of Asia (Thailand).

In the rice-growing village under consideration, the practice was to await the first rains in the spring before plowing the land with a simple point stick. If the first rains were late, the rice did not get sufficient start to keep its head above the later and greater floods. It would be advantageous to plant the first rice at the proper time in seed beds watered by irrigation and when it had a good start, to transplant it after the first rains. The extension service in that country had been trying to get the folks to break the heavy land early with a steel-shared plow, so that planting could be done with the first light rains. However, no one individual could do those things because village custom permitted the animals to wander about the tillable land until the whole village set about planting. Even though lands were in private pieces, here the individual could not go ahead.

However, this individualism of the Euro-American rural dweller has its disadvantages from the extension point of view. Obviously many extension practices cannot work unless adopted by the group. Irrigation, public works, much of soil conservation, and a great deal of the improvement in crops and animals depend upon the whole community doing the same thing at the same time. Illustrations are numerous. It takes a community to finance some of the heavier types of machinery needed for soil conservation. One man can seldom do it alone, unless he is extremely wealthy. In our Southern states it is often difficult to secure standard lint lengths on cotton unless the whole community adopts the same type of seed. Cross fertilization with indifferent seed in a neighbor's plot will ruin the individual effort to improve the cotton variety or yield. It is necessary for the whole community to act together.

In contrast to these types of practices, whose adoption is delayed by the individualism of the Euro-American, are those practices which are speeded up because of the extreme commercialization and division of labor in Euro-American society. When the seed houses put out an improved type of cucumber or tomato the very practice of purchasing rather than raising seed leads to changes which sweep a nation in a year or so. The phenomenal spread of hybrid corn is a good example; [5] in seven years it replaced open pollinated varieties in the American corn belt.

In southeast Thailand a fungus destroyed the pepper industry because community practice held to the same old ways; whereas, the sugar crop of Cuba was saved from destruction by a similar fungus because the planters by acting together practically all adopted a fungus-resisting cane, imported and bred quickly from a disease-resisting strain found in Java.

The individualism of the Euro-American rural dweller leads him to imitate the city in his family practices. His family is reduced to the size of the city family. This is an asset from some points of view, because funds and energy formerly devoted to the care of the larger family can now be devoted to improved agriculture. The farmer has money for the new seed or improved stock. On the other hand, the decreased size of the family works against the adoption of new practices suggested by the extension worker in that it adds to labor costs. Further, home labor is available more often the year around, and the large family can make many improvements during the time when crops do not require the whole energy.

The Euro-American rural dweller is one then who differs from the "folk" type of rural dweller [6] in that:

He produces primarily for the market.

He is socially close to the city.

He responds more rapidly and rationally to the city and its values.

He uses the same system of measurement of values and quantities as does the city.

He is aware of urban mentality and systems of meaning.

His economic desires are similar to those of urban Western society.

He is imbued with a value system known as "parity" in that he desires standards of living similar to those of the urban dweller.

He is a person with a residence in the country, and land and soil have no great sacredness to him.

His relation to the land is similar to that of an entrepreneur to an economic factor of production.

[5] Dept. of Agriculture, *Yearbook*, 1940, pp. 518-19.

[6] A good deal of data on this subject will be found in P. A. Sorokin, C. C. Zimmerman, and C. J Galpin, *Systematic Source Book in Rural Sociology* (3 vols., Minneapolis, Minn., 1929-1932).

He dissipates his soil resources if there is no immediate gain in conserving them.

He is a person accountable only to the general public, so he can act freely from the economic point of view.

His relationship to land is similar to that of personal property, in that land can be transferred easily from person to person.

He is handicapped to some extent by the weaknesses of local groups like the family, the clan, or the common village system.

EXCEPTIONS WITHIN EURO-AMERICAN CULTURE. The above is an overall picture which must be subjected to many qualifications and exceptions. The qualifications are general, political, regional, economic, and racial.

The general qualifications are as follows. With relatively few exceptions, rural life and its professions are still ruralized even in the most advanced or Westernized sections of Euro-American civilization. The country person still continues to produce some items for himself and for home consumption, even in the most commercialized agriculture. While there are farms where the operator produces nothing but potatoes, lettuce, grain or fruit for sale, and nothing else, these are a small minority—a small part of one per cent. Rural life tends to produce within itself a coherent system of values in which each element is necessary.[7] This constant factor in rural life is based on all the permanent characteristics which differentiate the city from the country. They have been set forth clearly elsewhere and need not be repeated here.[8]

Whereas the Euro-American rural dweller is socially close to the city, he still is a country person. He does not accept all the city at its face value. He selects from its city and adds to his own. He is the last to participate in some social movements in the city and the first to rebel against others. The land, the labor, the closeness to nature, the reality with which he lives, the animals and living things, the crops,

[7] Social systems tend to move one way or another constantly unless changed by "historical accidents." Farming, as a social system, tends to develop a complete agrarian system of values different from urbanism and does so to a limited extent even when constantly broken up by urbanism.

[8] See P. A. Sorokin and Carle C. Zimmerman, *Principles of Rural-Urban Sociology* (New York, 1929) Chapter II, and *The Systematic Source Book for Rural Sociology* (Minneapolis, Minn., 1930), Vol. I, Chapter VI *et seq.* To a considerable extent their ever-present influence on rural life makes understandable the Paratian conception of "constants" (residues) and Sorokin's ideas of logico-meaningful integration of a culture.

the seasons, the weather—all these have an influence upon him. He cannot help but notice that the land which he takes care of also takes care of him. He still is a local person subject to the mores and opinions held in common with his neighbors and other local persons. He still has a feeling of the oneness of country people and of their common interests in opposition to those of the consumers in the city. He sees his neighbors doing the same things as he does and this gives him a feeling of greater closeness to them. He may not trust his neighbor on the next farm but he has a deep insight into his mental make-up.

The differences between the city man and the Euro-American country man vary from region to region and according to types of farming. The fundamental character of the difference is that the Euro-American country man is more like the folk peoples of Asia than is the city man. Many of the same fundamental approaches to the folk of Asia will work, at least in part, for the Euro-American country man.

Political qualification of the previous outline of the character of the Euro-American rural dweller varies for different countries. Nearly all the sovereign states in the Euro-American part of the globe have legislated widely concerning the status of most country dwellers since 1914.[9] Much of this legislation has varied in intent from country to country, but all of it is similar in one respect. No matter how differing the objectives of recent agrarian legislation in Russia, Germany, Italy, England, and America, all of it has picked out farmers or country people as a class and has legislated for them or against them as a class as opposed to the general public. Much of this legislation was politically necessary for different reasons. In the democratic countries, it was necessary to meet the demands for "parity." In the totalitarian countries, it was considered necessary to achieve the aims of governmental regime, no matter whether these aims were good or bad. However, all were similar in the respect that legislation applying specifically to farmers was used in all the countries; and to the extent that the legislation was put into effect, the farmers and country people came to think of themselves as a group more like each other and differentiated from the city.

The probable net effect of this legislation was to increase the folk characteristics of the country people, to integrate them as a unit as

---

[9] See N. S. Timasheff, "Structural Changes in Soviet Russia," *Rural Sociology*, II (March, 1937), pp. 10 ff.

against the other social classes. Labor legislation requiring collective bargaining and other measures calling for concerted action has served to increase the cohesiveness of labor groups; so farm legislation of a similar nature tends to increase the cohesiveness of country people. Cohesiveness of country people and their feeling of oneness in opposition to all nonfarming or noncountry people is to some extent a folk or agrarian, solidary characteristic and works against the specific "urbaneness" of Euro-American culture outlined in the first part of this chapter.

This is a delicate distinction and one which has many fluctuating currents. Agrarian legislation (put through by farm blocs), which tend to increase the cash incomes of farmers and to increase their commercialization, made them more like city people from the standpoint of being purchasers of the same articles used by urban consumers. On the other hand, it gave them a common feeling in opposition to the city and to other groups because of the collective methods used in achieving the legislation and carrying out the common actions made possible by these laws.

RURAL AREAS IN EUROPE. We know only a little about what has really taken place in some of the European countries, and the legislation in all of them has been affected by the war situation. Nevertheless, here is a brief résumé of probable results in some of the more important countries.[10]

Russia has attempted to increase collective farming since 1929. The first five years reduced 18,000,000 small holdings to a quarter of a million big holdings in which the attempt was to make each village a collective farm. To some extent the resultant social organization resembles the *Mir*, or collective village, of old Russia.[11] New elements in the material culture include machine farming, central tractor stations, and so on.[12] No one knows what will be the end results of these enormous changes in the life of the people. Nevertheless, there are strong reasons to believe that legal collectivization, destruction by warfare, military government and resumption of Russian authority

[10] See Doreen Warriner, *Economics of Peasant Farming* (New York, 1939), for a résumé of the European situation.

[11] Timasheff, "Structural Changes in Soviet Russia": "There has probably *never* been a more agitated history of structural changes in rural areas than that of Russia during recent years." (Italics supplied)

[12] P. A. Sorokin, *Russia and the United States* (New York, 1944), pp. 204 ff.

over occupied areas will all have the net effect of increasing the col-lective psychology of the rural people and of decreasing the former economic individualism. The people will probably cling together in a general opposition to outside influences. However, to the extent that a new population is settled by the victorious Russians in rich sections occupied by the enemy and depopulated by the ravages of war almost anything may happen.

Germany attempted in the early years of national socialism to re-create in that country a new form of collective farming somewhat similar to an agrarian rural feudal class.[13] The Nazi revolution cloaked its rural plans with a splurge of dreamy idealism. Just what was really accomplished we do not know. At any rate, the matter is unimportant because the movement soon showed its teeth as pan-Germanism and the ensuing wars left the farmer more or less free to write his own ticket as long as this coincided with the food needs of a hungry people in time of war. However, the story is not over yet, in that the Allies will undoubtedly occupy most of Germany and force upon it the type of regime the various military governments feel necessary. This probably will mean a decline in economic prosperity for the farmer there, because he will have to pay his share of the war damages col-lected. Eventually it may mean a return toward more collectivism and less individualism to the extent that the farmers by sticking to-gether will mistrust outsiders. Further, the depleted net incomes of these peoples will force them to greater reliance upon their own natural resources.

What has happened, or will happen in the countries of Western Europe occupied by the Germans, only time will tell.

In England and America [14] the farmers have been less affected by agrarian legislation than have those of the Continent proper. Russia, Germany, Italy, and most other European nations changed agricul-tural social organization drastically. Most of this legislation has been an attempt at least to increase the real income of the rural people, and to the extent that it has been effective, it has increased their individ-ualism and their similarity to city people. However, collective legis-

[13] See John B. Holt, *German Agricultural Policy, 1918-1934* (Chapel Hill, N.C., 1936).
[14] See Carl T. Schmidt, *American Farmers in the World Crisis* (New York, 1941); Chester C. Davis, "The Development of Agricultural Policy since the End of the World War," Dept. of Agriculture, *Yearbook*, 1940, pp. 297 ff.

lation does make rural people more conscious of themselves and their opposition to consumers as a class. The end of the war will probably not bring an immediate increase in individualism.

Regional variations in the economic rationalization of the Euro-American farmer are of great importance. In general, eastern Europe is much more collective and less rationalized than western Europe. However, there are exceptions in that some of the large grain-farming regions of eastern Europe are more dependent upon cash income and many of the mountain villages of the western and southwestern part are still folklike and collectivized. Northern Europe is in general more individualized than southern Europe. Many of the sections of the latter—Italy, France, Spain—still keep the medieval village forms and are inclined to look inward upon their own production-consumption to a greater degree than do the farmers of the northern plains. However, there are many local exceptions. The English farmer before the war was much more of a producer for the market and a buyer of urban and manufactured goods than was the average farmer of the Continent.

In the United States, the North and the West certainly produced more of the market type of farmer than did the South.[15] The cities are largely in the North, and the best lands for commercial farming are there. However, the chief differentiating factor in the United States is the presence in the South of the Negro and the hillbilly. The "hillbilly" is the old American pre-Civil War type of farmer, the descendant of the "Sons of Liberty" who formed the radical element in the Revolution and Jacksonian movements.[16]

Economic factors are chiefly based upon the material wealth of the country person. All through Euro-American culture, the wealthy farmer, the big landowner, the landlord, and the commercial farmer is much more of the Euro-American individualistic and rationalized

[15] See Carle C. Zimmerman and Nathan L. Whetten, *Rural Families on Relief* (W.P.A. Research Monograph XVII, Washington, D.C., 1933).

[16] For the emergence of this class see A. M. Schlesinger, *Colonial Merchants and the American Revolution 1763-1776* (New York, 1918); Irving Mark, *Agrarian Conflict in Colonial New York, 1711-1775* (New York, 1940); C. A. Barker, *The Background of the Revolution in Maryland* (New York, 1940); Charles A. Beard, *Economic Origins of Jeffersonian Democracy* (New York, 1915); O. G. Libby, "Geographic Distribution of Votes on the Federal Constitution, 1787-1788" *Bulletin of the University of Wisconsin,* Vol. I (1897).

type than is the small farmer. Many of the regional variations depend upon the distribution of types of soil and topography which make large or small farming feasible.

Not only in America, but also in Europe, and wherever Anglo-American culture has spread, a small proportion of well-to-do farmers control most of the best land and economic resources. This is also true in other cultures, to at least a limited extent. In America, 20 to 30 per cent of the farmers control most of the lands good for commercial farming, and these families are the ones who get the greater proportion of the cash income received by rural people. They produce for and sell in the market and naturally they buy there. Hence they tend to be economic men, entrepreneurs, the citylike type of rural people, those holding most dear the traits attributed as being relatively more potent among the Euro-American countryman.

This differentiation in the United States is shown by the two groups interested in agricultural legislation.[17] One is the Farm Bureau which represents the commercial farmer. The other group, interested in the small-scale farmer (in many respects similar to his prototype in a folk society), is represented by the forces behind the Farm Security Administration. The farmer high on the rural economic scale wants higher prices, parity, agricultural adjustment. The lower economic group, the majority in numbers, wants the things dealt with under the farm security title. These last range from teaching the proper methods of home canning to loans and guidance during the crop growing period, or to locating on a little place of their own.

The same applies to Europe. In Russia the large farmer is the Kulak, in Germany the Junker, in England the Farmer or the Squire. All these classes, commercial farmer, kulak, junker, farmer, and squire, are economically highly rationalized, and to them our characterization of the Euro-American farmer most truly applies.

Mention should also be made of the village or nonagriculturally employed "rural" persons in Euro-American culture. Their problems are peculiar in one sense of the word, and similar in others. They work for farmers, sell goods to the farm people, help process farm products or are engaged in rural industries or services. Nearly all of them have access to small pieces of ground for supplementing their incomes with gardening and subsistence activities. During periods of national stress, such as in wartime, these are used to considerable advantage. Witness

[17] See John D. Black, *Parity*, p. 11 *et passim*.

the present Victory Garden movement in the Euro-American countries and the enormous amounts of supplementary food produced. In practically all the countries the problem is the same—teaching the village dweller to want to produce supplementary foods and the best methods for production and conservation. In continental Europe he is more important as a "part-time farmer," family members helping with the work.

Race as a differential among country people in the Euro-American cultural milieu is primarily important in the differences between the situation of the whites and the Negroes in the Southern states of the United States. The reasons are probably numerous and of ancient origin.[18] Various theories have been advanced to explain the differences, but for the problem of extension work, it is sufficient for us to know they exist. The Southern Negro constitutes a separate caste and lives under a decidedly different social regime from the white. On the whole, his income is small.

The whites in similar occupations do not fare much better. A large number are share croppers on the plantations—one-year tenants with considerable mobility from year to year. They do much of the work of cotton producing. Their homesteads are little developed. A lack of education, disease, high credit costs under the time-merchant system, and concentration on producing the cash crops with their variable returns characterize them. What money they do make is spent for urban or processed goods. But since they make little and have low standards of living, most of their income goes for food and the bare necessities of life. The fact that they are a separate caste makes it axiomatic that they must have their own educational leaders. But the leader is handicapped by the fact that the somewhat similar situation of the Negro is a cumulative and desperate one. Further, the ruling group, and most often the landowner and credit merchant, are white. The Negro's way of doing things resembles a folk society but his family is not the strong unit found among true folk peoples. On the economic level the Negro is a spender, but he seldom has anything to spend. He produces very little for home consumption, contrasted with the small farmer in other regions. In any agrarian program the American Negro

18 See Paul H. Buck, *The Road to Reunion, 1865–1900* (Boston, 1937), Chapters VI and XII; Benjamin Brawley, *A Social History of the American Negro* (New York, 1921), Chapter XIV; T. J. Woofter, *Landlord and Tenant on the Cotton Plantation* (WPA Research Monograph No. V, Washington, D.C., 1936); Charles S. Johnson, *The Negro in American Civilization* (New York, 1930), Chapter IX.

deserves and must have particular consideration. General rules applying to Western culture do not apply fully to the Negro because of his situation.

CONCLUSION. The Euro-American farmer or rural person has many of the social characteristics of an "economic individual" and will react to extension work motivated to a considerable extent by calculations of profit and loss. However, considerable variations exist according to region, type of farming, and economic and social conditions. This farmer or rural person has some of the traits of a folk person, and can also be taught to improve his farm practices and living conditions by the welfare approach. Much improvement can be made by suggestions along noneconomic lines such as "Grow vegetables for vitamins," "Paint your house for beauty," or, "Clean up the surroundings to prevent sickness." Good judgment and experience will be necessary for the proper emphasis in extension.

Chapter 11 · EXTENSION IN THE UNITED KINGDOM · By Robert Rae

BEFORE ATTEMPTING TO DESCRIBE the development of the Extension Service in the United Kingdom it may be desirable to outline very briefly the role that the extension service is required to carry out.

It will be agreed that agricultural science has, particularly during this century, already made considerable contributions toward a furtherance of increased, improved, and more efficient agricultural production. It will also be agreed that if the existing knowledge made available by the plant breeder, the chemist, the workers in nutrition genetics, veterinary science, control of pests and diseases, the engineer, the farm economist and all the others, was put into universal practice then there would be an enormous and immediate rise in production. The truth, however, is that not even in the most highly organized countries has all this knowledge been put into practice by more than a much too small proportion of farmers. While the application of existing knowledge would result in an immediate rise in efficiency on all farm holdings, for continuing efficiency the attainment of new knowledge is equally important. Scientific research is the fountain of industrial progress. Adequate provision for agricultural education and research must therefore be an essential part of the policy of any country which hopes to develop and maintain its agriculture.

The accumulation of new knowledge of how to grow bigger and better crops, how to produce and maintain more efficient livestock, must be of only very limited value unless it can be put into general use. The problem therefore is how to devise the best methods of "getting across" to the farmers the findings of the research workers in their laboratories or on the farms attached to research institutions and agricultural colleges. In most countries with an organized system this is the work of the extension service, although it is known under different names in different countries.

In those countries with an advanced agriculture and an established system of extension workers the main problem is to get the existing knowledge put into practice by all the farmers or, in other words, to raise the level of efficiency of the mass to something more nearly approaching that of the top ten per cent. To achieve anything like such a degree of perfection will call for a large increase in the personnel of the extension service.

HISTORICAL SURVEY. Until toward the end of the last century in the United Kingdom agricultural education and research, on which extension services must be based, was left to private enterprise. It was not until the Board of Agriculture was set up in 1889 that any State assistance was given to provide instruction for the farming community. Up to that time there was one institution for education, the Royal Agricultural College at Cirencester, and one institution devoted to agricultural research, the Rothamsted Experimental Station founded in 1843 by Sir John Lawes, who financed and later endowed it. There were some other private places of education, and occasional courses of lectures were given at the Oxford and Edinburgh universities. The Royal Agricultural Society maintained an experimental station at Waborn which worked closely in touch with Rothamsted and also conducted an annual examination and granted a diploma in agricultural and the allied sciences.

In 1890 an Act of Parliament empowered County Councils to spend on technical instruction certain grants made to them from the National Exchequer. Those grants, added to the contributions from the County Councils, enabled classes and courses of lectures to be initiated, and in addition many counties combined to form agricultural colleges or set up departments of agriculture within the existing universities. Most of the now well-known agricultural education institutions were set up in the early years following the passing of the Act. In the beginning, farmers were indifferent, sometimes skeptical, of these new ventures and "college trained" men, but gradually the colleges established themselves as educational training institutions and also acted as centers for lectures, classes, and demonstrational work in their respective areas.

The next important step was the Development Fund Act of 1909. By this Act a considerable sum of money was set aside and among the specified purposes for which it could be used agricultural education and agricultural research were included. Previous to this Act research had not been given any direct State financial assistance. Most of the national research institutions owe their creation to the passing of this Act. One other major result was the initiation of the Farm Institute scheme to supplement the colleges. These farm institutes were controlled by the County Councils but the Board (now Ministry) of Agriculture made grants of up to 75 per cent of the capital expenditure necessary to establish and equip them. The Board also repaid to the Councils two thirds of the annual expenditure necessary to maintain them. Although minor alterations have been made from time to time, this basis of financial responsibility still remains.

THE EXTENSION SERVICE UP TO THE OUTBREAK OF WORLD WAR II. The term "Extension Service" as applied in the United States was not used in the United Kingdom, but the same general field of work was carried out by two branches of workers in agricultural education, known as the County Agricultural Service and the Agricultural Advisory Service. For advisory purposes the country was divided into a number of advisory provinces with the central headquarters generally situated at an agricultural college or a university department of agriculture. At these centers were stationed a number of specialist officers who acted as advisors in their own subjects—entomology, mycology, bacteriology, agricultural chemistry, animal health, agricultural economics, and occasionally other subjects. The advisory officers acted as consultants to the county agricultural staffs and the cost of the service was borne by the Ministry of Agriculture. The second section was the County Agricultural Service and the chief officer was known as the Agricultural Organiser.

The size of individual counties varied considerably, and the number of farm holdings ran from about 3,000 in the small counties to 20,000 in the large counties. If the total number of holdings was divided by the number of counties the average number of holdings per county was just over 6,000. Many of these holdings were very small and only provided part time employment. If these were eliminated the average number of effective farm holdings per county would not exceed 4,000. In a typical county the staff was expanded to include one or more general assistants in the case of the larger counties, and specialist assistants in dairying, horticulture, poultry, and occasionally in other subjects. Up to the outbreak of the present war some twenty-five farm institutes had been established.

The County Agricultural Service was a local service in that the officers were appointed by, and servants of, a committee of their own County Council. In some instances, this committee was the Agricultural Committee of the County Council which had responsibilities in addition to agricultural education, and in other cases a special Agricultural Education Committee was formed. Those committees were preponderantly, although not exclusively, comprised of landowners and farmers. In those counties in which a farm institute was located, the county organizer was generally also the principal of the institute, which acted as the headquarters for the county staff. In the larger counties, district officers were situated in convenient market towns in their own portion of the county. In those counties without farm institutes the headquarters of the county staff was generally situated in the capital town of the county together with the other offices of the County Council.

The typical farm institute contained dormitory accommodation for from forty to sixty students, the necessary classrooms and work-shops, farm buildings, and land (running to some hundred acres). The function of those institutes was to provide instruction for boys and girls from the age of 16 upwards. The duration of the course was one year and the instruction consisted of indoor classes and lectures combined with practical demonstrations and work on the farm departments. The syllabus was designed to fit the requirements of young people who were going into practical farming, the majority of whom were the sons and daughters of farmers and farm workers. When the Corn Production Acts were repealed after the end of World War I, and as some solatium for the withdrawal of the subsidies on the growth of wheat and oats, the government appropriated a further sum of money to be spent over a period of some years for education and research, and at the same time made financial provision to award scholarships for the sons and daughters of agricultural workers to enable them to attend farm institutes and agricultural colleges. The scholarships covered all the expenses of such students and were instrumental in providing opportunities for boys and girls who otherwise could not have attended such institutions.

The farms attached to the institutes were run on good, sound farming lines, but their main function was to act as demonstration units where the results of research work could be tried out and tested under local conditions. These demonstrations included feeding trials, fertilizer trials, crop variety trials, machinery demonstrations, and so on. In addition, experiments dealing with a particular problem in crop or animal husbandry in progress at a research center could be carried out on a cooperative basis at a number of institute farms, with results obtained quickly on a wide basis and under a variety of conditions. Those demonstrations and experiments at a county farm institute not only provided a useful addition and stimulus to the students' training but also made the institute a central meeting ground for groups of farmers to meet, see the work in progress, and discuss its applicability to their own farms. The institutes also acted as a center for discussion societies which proved of great value as a method of securing the cooperation and interest of the farmers. They also served as the headquarters for many other organizations such as Young Farmers' Clubs.

The body responsible for the development of extension work within a county was a committee of the County Council, largely composed of farmers. There were other committees of the County Council responsible for other agricultural activities such as animal health, which also helped to spread over a wider group the responsibility of farmers for the management and control of schemes designed for the improvement of their own industry. There were in addition many other societies or associations organized in the main by farmers themselves for a great variety of purposes. Many of these were in a wide sense educational, in that they were directed towards technical improvement in one direction or another and the agricultural organizer was generally associated with all such activities. Talks, lectures, and the showing of educational films all provide a media for passing on information, and these were conducted whenever possible through some existing organization such as the National Farmers' Union. In many areas discussion groups had been formed which have been meeting regularly for a number of years. Such groups were much more useful than the single lecture in that they lived up to their name and became truly places where discussion rather than merely sitting and listening to a talk took place. It was not always the most able farmer or the one with the most valuable information to impart who was most ready to talk, but where a discussion group had been meeting regularly for some considerable time much of this feeling of shyness had been overcome.

In addition to the provision of technical advice to farmers, the extension service was also concerned with many other activities. Chief among these perhaps was work among the rising generation of farmers and farmers' wives. Such work was largely organized through the National Federation of Young Farmers' Clubs. Similar movements operated in many countries under various names and are so well known that no description of them is necessary here. They offer, however, one of the main hopes of spreading the extension service over a much wider range of farmers than has prevailed in the past. These young people are in their most receptive stage when they can be trained to keep in touch with and make full use of technical developments. The value of such clubs however is not limited to the technical field. They offer an opportunity to develop a community spirit and all that such a phrase should imply. If the clubs make meetings and discussions a major part of their program, opportunities are given to these young boys and girls, the men and women of the future, to develop leadership, to learn how to organize and conduct meetings, how to marshal their thoughts and how to stand up on their feet and express them. It would be difficult to overemphasize the importance of such movements in the general agricultural well-being of any country.

The womenfolk of the farms and villages had their own organization known as the Women's Institute. This movement has now been in operation for a considerable number of years and has spread widely over the countryside. It was organized into County branches with a national federation. The local women's institute, often with a building of its own as headquarters or making use of the village hall or school, held regular meetings throughout the year. The activities covered a wide range of interests such as lectures and discussions, music and drama, rural industries and handicrafts, fruit bottling and preserving, cakemaking and many others. These institutes served as a common meeting ground for the women of the countryside and did much to enrich and improve the amenities of rural life. That section of the extension service concerned with this type of work did much to help in the development of the women's institutes.

Developments during World War II. The United Kingdom is a food-importing country. Prior to the outbreak of war, home agriculture produced only about one third of the national food requirements. The difficulties and dangers of shipping and the necessity of diverting shipping to the carrying of other war requirements meant that an immediate and intensive drive for greatly increased home production of food was necessary. To carry out this program a War Agricultural Executive Committee was set up for each county and was responsible for the execution of the Government's agricultural policy. The membership of these executive committees was small, generally seven, of whom five were farmers or landowners. To carry out their work they appointed numerous district and local committees comprised entirely of farmers and technical officers. Among the committees set up by each County Executive Committee was one called the Technical Development Subcommittee, which took over, for the period of the war, all the duties of the extension service and therefore became responsible for all the technical help and advice given to farmers.

It may be of interest to note in passing that this method of organizing the wartime agricultural program has resulted in a large body of farmers throughout the whole country obtaining a considerable experience of the planning and execution of programs of work and of contact with their fellow farmers both on an executive and technical level.

One of the early duties given to the War Agricultural Executive Committee was to carry out a survey of all the farms in their respective counties. Farmers were classified into grades according to the degree of efficiency in the management of their farms, having regard to the type of land and other environmental conditions. It was the responsibility of the committees to do everything in their power to raise the standard of efficiency of those farmers placed in the lower grades. This work was carried out largely by the technical officers of the Technical Development Subcommittees. These officers were the previous extension workers augmented to meet wartime needs. The effect of all this was a concentration of technical effort directed towards those farmers who most required technical help. Their farms were visited at regular intervals and, as the wartime necessity required the maximum degree of efficiency from all the departments of the farm, both crops and livestock, there has been a steady and sustained improvement in production.

It was impossible, with the limited staff available, to visit all other farms with the same degree of regularity or frequency. Many other methods of imparting and disseminating technical instruction were used. For example one interesting development has come to be known as "farm walks." This consisted of a small group of farmers visiting, generally in the evenings, a good farm in their own neighborhood and walking over it with the farmer and the technical officer. There might be experimental or demonstration plots to see, but the whole system of that particular farm was examined and discussed. It is not always possible to persuade an outstanding farmer to address a formal meeting, but if he is at all public spirited he is generally quite willing to talk and discuss agricultural matters on his own farm. It can stand repetition, that the more programs operated by and through farmers themselves the better should be the results achieved.

Farming is a business undertaking and like any other its results must eventually be translated into terms of money. In some areas local discussion groups, especially where they consisted of men farming similar types of land and sizes of holdings, evolved a system of farm production records. A record sheet was devised on which each farmer set down under the requisite headings a summary of the purchases and sales for his farm for the month. These statements were sent in at monthly intervals to the District Office of the War Agricultural Executive Committee under a code number. At the end of the year the results of the year's farming were worked out for each farm and expressed in a standardized tabulation to show production efficiency and the percentages of income derived from the various commodities. The summarized annual statement indicated each farmer's position under his own code number only, but he could compare his position in relation to all the others given. Discussion of the results gave rise to very great interest and the first step to improvement in efficiency may often by a realization by the individual of exactly his position in relation to the efficiency of his neighbours.

Opportunities for extending advisory work may often present themselves when an improvement in marketing or some other process is being made. In the middle period of the war a National Milk Testing and Advisory Scheme was introduced. Under this scheme all milk produced was liable to be sampled at the first point of delivery— creamery, wholesalers' premises or whatever it may be. The sample was examined bacteriologically and on this basis graded into three categories. The names of producers whose milk was unsatisfactory were immediately notified to the County Agricultural Executive Committee, and as soon as possible one of their technical officers visited the farm to investigate the cause. In this way the farmer was not simply informed that his milk was bacteriologically unsatisfactory but was given technical help to find out the reason. The keen outstanding type of farmer will make himself familiar with new techniques arising from research work, and when problems arise on his own farm knows whom to call in for help and advice. For over-all improvements in any country, however, it is necessary to see that technical help gets to the less efficient farmers.

It is estimated that in the United Kingdom at present, instead of one farmer in eight who was probably in close touch with technical facilities before the war, seven out of eight are now seeking technical advice. While much of the wartime control of farming will undoubtedly disappear after the war, the extension service has had an unique opportunity during that period to get into intimate touch with a much larger proportion of the farmers than ever before. Everything possible should be done to preserve this contact to as great an extent as possible.

Broadcasting and films provide valuable means of imparting information to large numbers of people quickly and in an attractive form. Technical developments in both these fields have been very great during the war which, if they are properly used, should render them even more useful adjuncts to education in the postwar world. In the United Kingdom the agricultural press is a national press with a fairly wide circulation and a reasonably generous proportion of space allocated to educational or technical articles of a popular nature. The larger agricultural societies issue annual journals, largely devoted to the publication of technical articles and the results of experimental work. Most of the livestock-breeding societies also publish journals in which articles of technical value appear from time to time. Full advantage should be taken of all such opportunities for getting the right kind of information into the hands of the farmers.

Two further examples of the field of activity of the Extension service may be given. Both organizations existed before the outbreak of war, but wartime necessity gave a great stimulus to both. The first was the Allotment, or to use the wartime title, Victory Garden movement. Most Britishers have in them a love for a garden and one of the features of the countryside which most frequently impressed visitors was the gay little gardens found in every village. Where conditions in the towns did not permit of gardens for many of the homes, this need or urge for a garden was met to some extent by the formation of "Allotment" associations. An area of land was rented, as conveniently situated as possible, and plots were hired by individuals. The standard size of an allotment was normally 300 square yards or 10 rods. As a result of the wartime drive, when the number of allotment holders was added to the number who owned their own gardens, it was estimated that of the 12 million British families nearly 7 million were making a worth while contribution to the food supply. If the small 10-rod allotment was properly planned, cropped, and cultivated it should produce enough vegetables to last a family of five for eight months out of the year. Proficiency in growing vegetables, like anything else, has to be acquired by experience, often painful and expensive. Technical advice was given through the press, on the radio, by simple instructional films and by short leaflets often prepared on a cropping-plan basis. Britain was fortunate in the possession of a real radio star among its gardeners, a former extension service worker. His talk, "The Week in the Garden," every Sunday afternoon, had the largest audience of all the regular talks, even before the war. In the end, and especially when dealing with beginners, there is no substitute for practical instruction on the plot. Local associations of allotment holders were formed and extension workers in horticulture, local park superintendents and similar officials employed by local authorities gave a great amount of help to such associations.

The other organization, largely stimulated by the war, has been the formation of the Domestic Poultry Keepers Club for the backyard poultry keeper. The details of such clubs are not of importance in this discussion, but again it was the help and advice of the extension worker given by leaflet, over the radio, or by demonstrations that contributed much to their success. These poultry clubs and allotment associations admittedly owe much of their urge to wartime conditions, but they will continue into more normal conditions, especially the interest in gardening. The members of them will be drawn more largely from the towns than the country where such activities form part of the normal picture. In a highly industrialized country there is everything to be gained by more town people taking an interest in learning how things grow or are produced. They form a large part of the electorate and the more town and country people can appreciate each other's point of view the greater is the chance for the adoption of a sane and balanced agricultural policy.

POSTWAR DEVELOPMENTS. Plans have been announced for the unification in England of all branches of the extension service into one National Advisory Service. The service will be centrally controlled and financed entirely from the National Exchequer. Full details have not been made available, but there is no doubt that arrangements will be made to maintain local interest and guidance by the appointment of local advisory committees. The present structure of provinces and counties will be maintained, although there may be some boundary alterations. Without attempting to discuss all the implications of this change, two points may be mentioned. First, from the extension worker's point of view, the unification into a National Service should give a greater feeling of cohesion and cooperation, and presumably more opportunities for change and promotion. The second point is that, as all the expenditure is from the national purse and none from the county funds, the poorer counties financially, which were also in the main purely agricultural counties, should no longer be at a disadvantage as compared with their more wealthy industrial neighbors. With a nationally financed service the requirement for extension workers need no longer be limited to the financial standing of any individual county.

The second development which is envisaged is an increase in the number of farm institutes. It is intended that no county shall be without such facilities either within its own boundaries or in an adjoining county. This increase will allow a much larger number of the rising generation of farmers to have the opportunity of at least one year's residential technical instruction.

THE EXTENSION WORKER. It may not be inappropriate to conclude this chapter with a few comments on the importance of attracting the right kind of individual for extension work. Great care should be exercised in the selection of extension workers. Consider what is asked of him. He is asked at one and the same time to be practical farmer and scientist, guide, philosopher, and friend, able to talk the language of the scientist and translate it into the language of the farmer. He must know his district, his farmers, gain their interest and confidence, be prepared to work all day and many of his evenings, be able to assess the applicability or otherwise of new findings to his own area and develop his own machinery to get them put into practice. Like other professions there are good, outstanding, and doubtless indifferent extension workers. Their work in influencing the whole level of production in their areas is so important that there is no room for the indifferent type. When all this is expected of them, they are entitled to have the importance of their work recognized, and conditions of service and salary should be commensurate and they should not be expected single-handedly to achieve the impossible.

Agricultural education must be one integrated structure. The university departments or colleges must be staffed and equipped to train teachers, researchers, and extension workers. The research institutions must be staffed and equipped to deal with the investigation of existing problems and prosecution of new work. The extension workers must be sufficient in number and have the necessary facilities to carry out their portion of the joint endeavor. On them, in the end, must fall the major responsibility of having translated into farm practice, technical developments as they arise.

Chapter 12 · AGRICULTURAL EXTENSION SERVICES IN NORTHWEST EUROPE · By P. Lamartine Yates and L. A. H. Pieters

INTRODUCTION. The northwest corner of Europe [1] today is, in agriculture, technically more advanced than any in the world. It was the first off the mark in various fields of agricultural science. It cradled the agricultural revolution of the eighteenth century which replaced medieval farming by turnips and clover and by scientific livestock breeding. Later, these countries were the first to learn and practice the use of artificial fertilizers, which, in conjunction with favorable soil and climate, has produced yields per acre higher than in any other part of the world.

Socially the transition from medieval to modern society has been more painful and the result less completely satisfactory. The systems of land tenure which exist today result from the balance of political power during the period when the open fields were being enclosed, that is, from 1750 to 1850, the movement beginning earliest along the Atlantic and North Sea coast and being longest delayed in central and eastern Germany. France and Germany were the only two countries where large-scale landowners dominated the scene. But in France they were more interested in court life than in farming, and consequently when the French Revolution destroyed their social position, they made no further attempt to defend their position in agriculture, and the division of their estates among the peasants occurred without undue friction. Indeed, they were glad to sell for ready money to make good in part their lost fortunes.

In eastern and northeastern Germany the case was different. There the estate owners preferred farming to politics (Bismarck said he was more moved by a beetroot than by all the politics of the Continent); and they saw to it that the liberation of the serfs was carried out in such a way that most of the land which the laborers got had to be surrendered again to the estate owner in liquidation of debts. Thus there

[1] This chapter does not attempt to deal with all the countries of northwest Europe but only with six, France, Belgium, the Netherlands, Denmark, Germany, and Switzerland, of whose extension work the authors have firsthand acquaintance. The growth of extension services in these countries has been sufficiently various to provide illustrations of all the main types of service and of the chief problems still confronting those who are concerned for rural welfare.

grew up in this region a system of large farms worked by landless laborers who were paid very low wages, mostly in kind.

In the other four countries and in western Germany there were no large landowners. The farm land was owned either by the peasants themselves (though in Denmark this was achieved only late in the nineteenth century) or it was held in small lots by men of commerce in the cities, and leased to farm operators, this being typical of the Low Countries. The pattern of agricultural society which emerged in the twentieth century was, therefore, throughout this area except in the extreme east, one of small peasant farms, most of them owned by their operators, though some in some districts worked on a cash-tenancy basis. In France and Germany and to some extent in the Netherlands the rural population also contains large numbers of hired farm workers not owning land except perhaps a garden attached to the cottage. On the majority of farms only one or two hired men are employed and the status of the worker is more that of a personal servant or attendant than that of a factory worker or office employee. The ties are personal and the economic nexus is much less in evidence. No one is very "big" and no one very "small."

Another factor which has helped to bring this about is the land shortage. Throughout the whole of this area population density is great, and although industrial development has relieved pressure of population on the farm land, it still remains true that nearly every peasant in northwest Europe could cultivate and would like to cultivate more acres if he could get them. The typical size of farms is 15 to 25 acres—larger, of course, in some regions such as northeast France and northeast Germany, but even smaller in others such as Belgium and Switzerland. Without tractors and other expensive machinery most peasants could handle twice as much land. The very high land values prevent even successful peasants from accumulating much land at the expense of their neighbors, and where this does occur the process is reversed when the farmer dies and the property has to be divided equally among all his children. Consequently, there is a remarkable degree of social equality throughout the farm community. No one is markedly superior to his neighbor; even the hired laborer, if he is thrifty, may save enough to acquire a small piece of land and start farming on his own. This atmosphere has been especially favorable to the growth of cooperative societies for all kinds of purposes. There was no one section of the rural community too proud to collaborate.

It has led to a passionate feeling for political democracy, and it is perhaps significant that it has been in those parts of Germany where this rural equalitorianism was least well established that agricultural interests have supported the Nazi dictatorship. It did not, however, lead to political aggressiveness. The peasants had, by the end of the period of enclosures, obtained what they wanted and were suspicious of further change. Since then they have tended to support the parties of the right. The propaganda of social democrats, still less of communists, has made no appeal. Thus the human material with which the extension services of Northwest Europe has had to deal has been remarkably homogeneous, innately conservative, but not impervious to new ideas and not unwilling to cooperate for clearly defined and understood objectives.

THE EDUCATIONAL SYSTEM. There are three aspects to agricultural extension in northwest Europe: first is the system of general elementary education in rural areas; second, agricultural education; and third, advisory services provided for farmers. Although each of these is organically related to the others, it will be convenient to consider them separately, and although conditions vary quite appreciably from country to country, it seems better for the sake of brevity to attempt a general description roughly applicable to the whole area even at the risk of making some generalizations which do not fit the circumstances of the particular country. The reader who is personally intimate with the extension system in any one of these lands must therefore forgive the lack of detail and description of particular institutions.[2]

Elementary education throughout this area has been firmly established for a considerable number of years. In Belgium in 1900 there were admittedly still 20 per cent of the population who could neither read nor write, but today the proportion is less than 1 per cent in each of these countries. The school period lasts from the age of five or six up to thirteen in France or even to sixteen or seventeen in some of the Swiss cantons. In some of the rural areas, however, the last year or two of schooling is frequently dodged so that the child can be used for farm work. Moreover, in some districts schools close for two or three months during the harvest season. Whereas in some countries, as in the Netherlands, rural schools are kept at the same level as in cities with regard to staff, building, and equipment, in most rural areas the schools

[2] For a discussion, country by country, see P. Lamartine Yates, *Food Production in Western Europe* (London, 1940).

are less well provided for than in the cities: the buildings are old and inconvenient, equipment is lacking, there are too few teachers and the good ones are attracted elsewhere where teaching conditions are better. The results, it must be admitted, are disappointing. Children emerge from school able to do little more than read and write, with no desire for acquiring more knowledge, on the contrary, with a thorough distaste of learning. They do not perceive any relationship between the life inside the school and the life of their home, their farm, and their outside world. Very few children from these rural elementary schools go on to the ordinary secondary schools.

If agricultural extension work is to have greater success, then clearly the system of general education, which is the basic foundation, must be improved in many areas and in many respects. The first necessary reform is to raise the school-leaving age to fifteen in all these countries and to see that it is enforced. Along with this, there must be rebuilding of many of the village schools providing them with light, airy classrooms, modern sanitation, and modern school equipment. More teachers are needed and in many cases of better quality; this, having regard to the comparative isolation in which they are compelled to live, can only be secured by paying more generous salaries. There might well be more contact and consultation between teachers and parents and some attempt to link the education with the cultural traditions of the locality.

But when all this is done, it will not be enough. Children who leave school at fifteen, however well they may have been educated, are not yet fully equipped to become citizens in a democratic state. This is being increasingly recognized in regard to urban children. For instance, in several countries continuation education up to the age of eighteen is now compulsory in cities, but the same standards have not yet been applied to the rural population. There can be no justification for this neglect, though there is much justification for expecting that the needs of rural children must be satisfied by quite special methods appropriate to their environment. Ideas on this question take two forms in Western Europe: Some people advocate part-time continuation education analogous to that of the cities but with a different curriculum; others advocate the establishment of rural high schools for young persons aged eighteen or twenty who have already had four or five years practical work in the world since leaving elementary school. Under European conditions there is much to be said for the latter alternative, which of course owes its inspiration to the Danish People's High Schools. Among its advantages are that: 1) having already been out at work, the students have acquired a picture of conditions in the real world; 2) at the older age they are more likely to know what subjects they want to explore and why; 3) living together as boarders creates a firmer sense of community than the mere attendance of evening classes; 4) when they leave they are old enough to continue a certain amount of reading on their own and can take an intelligent interest in the problems of farming.

Whether the best solution is to be found in schools of this character or not, the important objective is to arouse an appetite for knowledge, a curiosity about the world, an interest in scientific method and in logical reasoning which can be applied to all sorts of different situations in later life. The Danes have come to possess qualities of this kind, and it is probably no exaggeration to say that the high general level of farming ability which prevails in Denmark, and the success of cooperation, results from the high general level of education, from the habit of inquiry, thinking, and reasoning which they have developed in their Folk High Schools. Unless human soil be well tilled in some such manner as this, the work of agricultural education and agricultural extension can bear little fruit.

AGRICULTURAL EDUCATION. It is a little difficult to describe in a few paragraphs the system of agricultural education, since there are substantial differences between the various countries of western Europe. All that can be done here is to describe a standard type which roughly corresponds to what most of them have, while asking the reader to remember that not all the countries have all the constituent parts and that, as regards the level of technical instruction, what may be called middle-school work in one country may be notably more advanced than work going under the same title in another country.

Broadly speaking, there is a three-decker system of agricultural education: first, a primary or elementary grade; second, the middle schools; and third, the colleges and university departments. The simplest form of primary education consists in evening classes given in the village school, usually by the village schoolmaster, on two or three days per week over a period stretching from autumn through winter into spring. In Belgium, for instance, the course is one of 100 hours in each of two winters; in France, 150 hours; and in the Netherlands,

200 hours. In the Netherlands about ten years ago elementary agricultural and horticultural schools were established. These schools are all maintained and administered by farmers' associations who for this purpose receive a subsidy from the government. These schools are training future farmers and farm laborers; the instruction is partly of a general character and partly technical, but adapted to local conditions. The instructor in charge has an official certificate for agricultural or horticultural teaching. The school is a continuation of the ordinary elementary school and the course takes four years. During the first year classes are given on two days a week and during the following years one day a week, so that most of the time the students can work on the farms and help their parents.

These schools seem quite successful in that they reach the sons of the small farmer and farm laborer, who would be unable to spend whole weeks at a distant school. An alternative type of primary agricultural instruction is given in what are called Winter Schools where classes are held for half the day or a whole day two to four times per week. In some countries the children proceed to these straight from elementary school; in others, for example Belgium and Germany, the entrance age is sixteen. Since the classes are held in the daytime, attendance is virtually impossible for hired agricultural laborers and for the poorer peasants' sons who cannot be spared from the farm.

Both these types of instruction are, in effect, a form of voluntary continuation education resembling the classes given at technical institutes in the towns. There is this important difference, however, namely, that in some countries a large part of the instruction is given not by specialists but by the regular village schoolmaster who is expected to have a working knowledge of the elements of botany, physics, chemistry, and zoology. In practice the instruction necessarily suffers from this: on the one hand it is often too amateurish; and on the other, being given by a person the children know so well, it lacks the freshness and authority which might attach to a stranger. In the Netherlands many village schoolmasters follow courses in agriculture or horticulture where instruction is given by the Government Agricultural Advisors. The teachers who successfully complete these courses are given a certificate which entitles them to conduct government-subsidized agricultural or horticultural courses.

A great part of agricultural progress in the Netherlands is due to the activities of these agricultural teachers. They generally remain close to the local problems and have a good understanding of the farmer's troubles. In the Winter Schools, however, some of the instruction is given by the county agricultural advisor (see below). Winter classes and Winter Schools in various aspects of home science are also provided for girls, but they are not always well attended, especially in France and Belgium, because most parents do not like their daughters going out alone in the evenings.

The second rung of the ladder is the agricultural middle school, which is either wholly a boardingschool or frequently attended both by boarders and day scholars. The level of instruction and the equipment of these schools varies greatly between the different countries. In France, for example, children may proceed straight to these schools from village elementary schools; in the Netherlands they must have previously attended a secondary school or, alternatively, one of the Winter Schools. In France most of the schools have a farm attached, and part of the training consists in practical work on the farm. In Belgium there is no farm and no practical work, since the farmers object to paying for their sons doing work which it is argued they could just as well do at home. In Belgium most of these schools are denominational, that is run by Catholic farmers associations, as are also some in the Netherlands. Elsewhere they are almost all State schools. The course extends over five or six winter months and lasts for two or three years. There may be some fifty to eighty students in each school. The subjects include physics, chemistry, zoology, botany, and biology. The schools have laboratories, not in all cases particularly well equipped.

Besides these middle schools for general agricultural education, there are also in some regions middle schools devoted to special subjects, for instance, to horticulture in the Netherlands, or in parts of France to viticulture. In addition there are home-science schools for girls, a considerable number in Belgium, Denmark, and the Netherlands, but some in each of the countries. Often the course for girls is given in the summer months when the boys' courses have been completed. This, however, does not encourage attendance, since the women are needed just as much as the men to help on the farms in summer.

The third and highest grade of agricultural education is provided in the colleges and university departments. The course varies according to country from three to five years. In Belgium and Germany the courses are mostly in general agriculture, whereas in the Netherlands and France there is considerable opportunity for specialization. The Dutch, for example, have at Deventer a college for tropical agriculture; the French have one at Rennes for dairying, at Grignon for crop husbandry, and at Montpellier for viticulture. In each of these countries, however, there is a senior agricultural institution (the Agricultural College at Wageningen and the Institute Agronomique at Paris) where degrees of full university standing can be obtained while the college final examinations only grant diplomas. In Denmark the highest institution is the Royal Agricultural College at Copenhagen and in Switzerland the Agricultural Department of the Technical College at Zurich.

This then is the hierarchy of agricultural instruction which has been built up in western Europe during the last seventy years. Its most elaborate organization is found in Belgium and the Netherlands, where substantially more money is spent per head of agricultural population than in France or Germany. It is said that something like half the male farm population of Belgium has at one time or another attended one of these courses or schools and that one sixth of the female farm population has had some instruction. In the other countries the coverage is far less complete, and in the French *departements* and the Swiss cantons where a large part of the funds has to be provided locally the local administration is not often willing to shoulder the burden of providing proper school facilities.

On the whole it has to be admitted that notwithstanding certain individual successes none of these different types of institutions has produced the results hoped for. Many people in these countries are anxious to find ways of doing better. One principal cause of trouble is the human material which enters these various agricultural schools. The children, most of them, lack even that minimum groundwork of general education that would enable them to understand these new subjects—they are too unformed, too inert to respond to technical instruction even if it were inspiring. Most of them have acquired a distaste for school and the last thing they wish for is to return to it again after their regular period of school is terminated. The remedy for this lies, of course, in the field of general education which we have already discussed.

Another trouble is the curriculum, which in many instances is too academic in character. In many of the middle schools the boys are given a thorough grounding in pure physics and chemistry without relating it sufficiently to practical farming problems. It might be better to start with the practical problems and only introduce pure science in order sufficiently to explain them. There is also an over-emphasis on examinations and on securing a career either as a civil servant or as a teacher. Few scholars from the middle schools and hardly any from colleges go back into agriculture and take up farming as their profession. This is partly a consequence of the teaching but it is also because the great majority of farms, however well managed, would be much too small to yield the kind of income which a moderately well-educated young man hopes to earn.

Finally, in some countries the teachers are, many of them, ill adapted to the work. Of those who teach in the middle schools and colleges each has himself had to go through the long years of technical training necessary to secure a degree in agriculture and has developed in consequence a professorial academic attitude to his subject so that he finds it difficult, often impossible, to put himself in the shoes of his listeners and explain things in ways which link up with their limited field of experience. Again, far too many of the teachers are expected to do half a dozen other jobs as well. In Belgium many of them are at the same time the provincial agricultural advisors carrying a host of administrative responsibilities, and in France, too, they are frequently the directors of the agricultural services in the *Departements,* responsible to the *prefect* for many administrative matters and also expected to develop contacts among farmers.

What then should be done to improve the quality and effectiveness of agricultural education? Assuming as one must, that funds will be limited, it would seem wise to spend such additional appropriations as may be obtained not on establishing more higher grade schools but on extending and improving those of the primary and middle grades. In these countries of small-scale farmers, agriculture cannot carry a large number of persons with university degrees in farming. What it does need is a large number of competent farmers who know how to get the most out of very limited resources. It follows that any extra money available should be spent on getting more instruction over to the would-be small farmer. This will mean bringing a new spirit into the schools, giving the instruction a much more practical slant, finding teachers who are less distant and more "of the people," giving more time to farm management questions and organizing visits to neighborhood farms. The object of the schools and their teachers should be to find out what the local farm boys need to know and base the instruction on these needs rather than on some preconceived idea of what they ought to know. A natural adjunct for the farm school should be a young farmers' club composed of young persons aged, say fifteen to twenty who undertake some practical task, such as rearing pigs or poultry, who organize for themselves lectures and visits and competitions and who hold frequent discussion meetings. These have proved themselves highly successful in a few regions of western Europe, and there is no reason why, given energetic leadership, they should not succeed equally well everywhere. They should act as a link between the formal education of the school and the day-to-day life of the lad working on the farm. Their existence should help reorientate the school toward the needs of the local community.

AGRICULTURAL EXTENSION SERVICE. In all these countries the technical education of the young is supplemented by an advisory service for adult farmers. The general pattern of this is much the same as in other countries of the Western World; that is to say, there is a team of specialists operating on the geographical basis of county or province whose responsibility it is to contact farmers in their area and help them with technical problems. Where the system is most highly developed there may be a team of as many as six or seven, including an entomologist, a mycologist, a dairy bacteriologist, a farm economist, and others. Where small funds are available there may be just one man with an assistant who has to deal with the whole range of problems that may arise.

In western Europe the system is organized in one of two ways. One system is for the whole service to be provided by the State, this being the case in France, the Netherlands, Germany, and to some extent Belgium and Switzerland. The other method is for it to be provided by the farmers themselves through their societies or cooperatives, this being the system in Denmark and to some extent in Belgium and Switzerland. In Denmark the farmers themselves contribute to pay the salary of their advisors and collectively have the right to appoint and dismiss them. This has the obvious advantage of getting what they want rather than what somebody else thinks they need. As a disadvantage may be mentioned the unquestionable lack of independence of the advisor, whose honest and well-considered opinion may run counter to preconceived ideas of interested influential farmers.

On the whole this system of advice whether by farmers or by the State is most fully developed in Denmark, the Netherlands, and Switzerland. In these countries the number of advisors is such that in the course of the year quite a high proportion of the farmers can be reached by personal contact, whereas in France, Belgium, and Germany the provision is much more parsimonious and only a comparatively small percentage of the farm population reaps any benefit.

One particular and most important aspect of the work has been the development of farm bookkeeping, which perhaps more than any other thing enables a farmer to understand his economic problems and by better management improve his income. This bookkeeping started before the first World War in Denmark and Switzerland. In Denmark it was undertaken by a large number of local bookkeeping societies, while in Switzerland it was centralized at the head office of the farmers' union at Brugg. Beginnings were also made in Germany but only among some of the large farmers who clubbed together to hire a professional accountant. Subsequently bookkeeping also developed in the Netherlands, organized by the farmers' societies, where just before the second World War there was a coverage of over 7,000 farms. In Belgium and France there has been no enthusiasm for this activity. It was not until the 1930s that any beginnings were made and even then progress was slow. Undoubtedly one reason was the fear that the bookkeeping results might get into the hands of the tax assessors. There is much evidence to suggest that, where bookkeeping is widespread, farm management reaches a high level and farming communities prosper. Experience also shows that hitherto at least the work only succeeds if freely undertaken by the farmers among themselves. Mention should be made of the establishment, several years ago, of cooperative management associations in the Netherlands, where a number of farmers employ a farm management expert to make a continuous study of their enterprise and give technological and economic advice for improvement of the farms. Generally the association will also take care of the bookkeeping of the farms.

Advice to farmers is provided in a number of other and more informal ways. The farmers' societies and cooperatives, the women's organizations, the various producers' associations, the animal breeding societies, young people's associations—all of these organize activities from time to time which directly or indirectly have some educative value. For example, in Switzerland associations of farmers' wives organize fruit canning in the villages and on the farms; and in Germany there are, or were, similar organizations for teaching cooking, hygiene, and other aspects of home welfare. In France the village *syndicat*—the local branch of one of the national farmer's unions—acts as a center for discussion of a multitude of local problems. In Belgium the *Boerenbond* provides an elaborate advisory service based on the village gild. In the Netherlands and Denmark the breed societies and the co-operatives disseminate much useful knowledge. All these various associations have the advantage of depending on the spontaneous support of their farmer members. Nearly all of them have grown up from below rather than being imposed from above. At the same time their purposes are not directly advisory. They are mainly concerned with specific functional jobs or with furthering farmers' political interests.

On the whole the extension services in western Europe work better than the formal agricultural education. Nevertheless in most of these countries there are far too few advisors, and those who are appointed are grievously overworked. In Belgium and France, for instance, the chief of the provincial or departmental advisory service is a man who besides having to care for some tens of thousands of farmers has to administer the Foods and Drugs Act, and the regulations for specifying the content of fertilizers, insecticides, and so on. He is responsible for testing weights and measures. He may have to administer State subsidies, for example, for the reconditioning of cowsheds; and he is usually expected to spend a substantial number of hours each week teaching in one of the agricultural schools. Naturally, he has not time to perform all these functions well. Furthermore he has not, in most cases, had training which would enable him to mix easily among farmers; he is often a professor or lawyer at heart, is too learned and too stiff in his behavior and conversation.

It is indeed questionable whether, even with the best possible organization and personnel, an extension service can function properly if it is wholly supported and administered by the State. The Danish experience suggests that it has to have its own grass roots and be based on the interest and support of the farmers. Development in the Netherlands points in the same direction. In Britain, where a State system has operated for many years, the wartime experience has shown what immensely improved results can be obtained when the farmers themselves are called upon to organize an advisory service and when in each district the best local farmers take on the duty of stimulating their neighbors by precept and example. This admittedly would be more difficult in countries where small farms predominate, partly because there may be large areas where no one has adopted modern methods and can be used as an example and still more because among those who are suitable there will be few who can spare sufficient time away from their own work to act as apostles in other parts of the district.

Nevertheless the attempt must be made. If extension services are to become more effective, they must be based on self-help in the villages. The smallest unit of agricultural organization should form the nucleus, whether it be a branch of the farmers' society or land workers' union or a cooperative. The work must be based on the needs of that particular village. It must use the village leaders. It must work in the village environment and demonstrate improvements on the most ordinary village farms, using the most ordinary tools and equipment. Farmers will believe the results achieved under conditions of their own making, while they remain skeptical of the results, however sensational, on State experimental farms where they suspect the reason to be unlimited availability of money. And there is another thing: it is true to say that throughout large parts of western Europe, particularly perhaps in France and Germany, the farmer has up to now been hostile to formal advice. He has resented being told what to do by an outsider. In order to make him receptive to new methods these have got to be put over to him by one who has gained his confidence through mutual understanding and who is thoroughly integrated in rural life, one of his own kind, one whom he respects but who farms just as he does and who shares the same social status. Knowledge can probably be disseminated with great rapidity if these channels are used, if the farmers feel they are finding out rather than being instructed, and if their leaders are of their own choosing.

CONCLUSIONS. This has been a very sketchy and inadequate survey of the educational and extension problems in western Europe. Deliberately, no attempt has been made to single out the special problems of the individual countries. That would have led into too much detail and obscured the main objective. The aim has been to show what the present system is and what are its limitations. When one uses the phrase "the present system," one is referring to the situation as it was in 1939. It is true that a number of years have passed since then, years of fighting enemy occupation and again fighting. But from such reports as are available it would appear that the general framework of the extension services has come through virtually unchanged, that therefore the system of 1939 will also be the immediate postwar system.

In considering making recommendations for changes three things have to be remembered: First, the very special social structure throughout the region with the strong emphasis on democratic equality, particularly in rural areas, expressed in the feeling that every man is good as any other, and including an intense dislike of orders and instructions from above. Secondly, there is the geographical environment. The region is a densely populated one, with few groups really isolated, and there is little of that group loyalty which isolation commonly engenders. Moreover, in some parts of the area the religious sanction no longer obtains, so that there is little by way of counterweight to the intense individualism which each peasant naturally has. Thirdly, each of these six countries has very special problems of its own, problems which differ radically from those even of other near-by countries, let alone of other continents. It would therefore be impertinent to suggest to these countries what reforms are necessary. Their peoples and governments are already perfectly aware of the questions discussed in this chapter. Hence these final paragraphs are addressed rather to readers in other countries who may find it instructive to ponder on some of the methods which Europe may use in overcoming the difficulties of extension work.

The central problem is, as has been said, to find some items of common interest which can be used as a focal point of joint activity within the village. Out of the village needs must grow the village reforms. That means that the order of priority will vary from district to district and almost from village to village. Some will start on cattle breeding, others on fertilizer experiments, others on land drainage, and so on. Having found the basis, the next step is to find the personnel, which assuredly must be recruited from among the people themselves so that they may tell the good news to each other. This means that one cannot propound a nationwide scheme outlined at government headquarters in all its detail. On the contrary it may be better to start with a few selected villages, one or two in each major agricultural region, and let the movement spread, if it will, from there. The State can, however, give help of an organizational and administrative character, provided it helps discreetly and remains as far as possible in the background.

It is not the purpose of this chapter to discuss in detail what kind of activities should be undertaken. So much will depend on local conditions. The sort of activities which come to mind are demonstration plots for fertilizers and pesticide experiments, cooperative machinery pools, development of bee keeping and poultry keeping, fruit and vegetable canning, a village dispensary, a village library, and village discussion groups, but there are a score more activities besides these.

Finally, opportunity may be found as the spirit of group loyalty develops to begin to give it an even wider orientation and link up the projects of village improvement with projects for whole regions or for the whole nation. The village might, for example, share in some national competitions with targets, say, for clean milk or for home canning or for village nurseries. A spirit of friendly rivalry between one community and another often provides a spur to even greater efforts and also helps the villagers, many of whom have little opportunity to travel, to picture themselves as members of a wider body and contributing in their small way to objectives willed by the nation as a whole.

Chapter 13 · AGRICULTURAL EXTENSION IN THE UNITED STATES · By Edmund deS. Brunner and C. B. Smith

AGRICULTURAL EXTENSION IN THE UNITED STATES deals with a population of about 30,000,000 persons located on 6,000,000 farms, the aggregate area of which is over one billion acres, and nearly 30,000,000 other people in rural areas but not on farms. Of these 15,000,000 rural families, extension work influenced 4,920,229 farm families and 3,289,-216 other families in 1943 alone, a total of over eight million.

Obviously, it faces a bewildering variety of conditions and situations. Some 200 crops engage the attention of its total constituency. There are areas where semi-tropical, and in some island territories even tropical, conditions prevail. In contrast there are areas where the growing season is as short as 120 days, or, as in Alaska, even less. The actual farmsteads vary tremendously. There are the huge tracts of cattlemen and corporation farms. The largest group, the family-sized farms, vary according to crops from 60 to 80 acres or even less, up to 160 and 320 acres or even larger. There are also the 10 to 20 acres of the share cropper in the Southern cotton belt, or the equally small holdings of the subsistence or part-time farmers. Farm management and organization are based on a single commercial crop here, on highly scientific diversification there, on production largely or wholly for home consumption elsewhere. Techniques vary from a high degree of mechanization to the intensive cultivation of small holdings by hand labor. Standards of living may be typified by the commodious, efficient, urbanized home of the citrus fruit grower; the spacious, comfortable but older homestead of the dominant type of household farmer; the neat bungalow set down on the suburban acres of the part-time farmer; or the wretched hovel or pioneer type cabin of the share cropper or mountaineer.

In racial origins there is comparable variety. Just as all peoples have contributed to the United States of America of 1945, so almost all groups are found on its land. Descendants of the Pilgrims and of John Smith's company are found in all forty-eight states, although still largely in the Eastern third of the nation. Northern European stock is comparably scattered but is disproportionately represented in the northern Mississippi valley. Here and there are settlements of Latin and southeastern European nations farming as intensively as in their mother lands. The Negro is indispensable to Southern agriculture, as are the Mexicans, and some whose forebears were Oriental, to sugar and the fruit and vegetable crops of the West. The original inhabitants, the Indians, also have some small stake in the land once wholly theirs. Hence there are communities dominated by the traditions and mores of a "foreign" culture. There are others into which many cultural streams have entered, some partly, some wholly blended into that new product of old cultures and new land and conditions—the American. Nor can any American rural community be wholly understood unless its cultural heritages are known.

Yet despite these and other equally important differences and contrasts, these six million farm homes are bound together by one dominant interest, the production of food or fiber for use. Be it the manager-executive of a corporation farm, the operator of a family-sized farm, or the week-end farmer on his suburban acres, all are concerned with tillage, with earth, and with the life they can build for their families out of the land and through the cycle of the seasons.

Moreover, despite these differences and with full emphasis upon the qualifications given, all these people display in their communities and socioeconomic organizations the characteristics of the Euro-American society described by Professor Zimmerman in Chapter XI. This means, in brief, that the production of food and fiber is primarily for sale, that rural life and organization takes on many of the aspects of urban society, that the land is viewed far more impersonally than in the Orient (a fact which accounts for the relatively high mobility among the farm population), and that the farmer and his family act independently of the village or clan.

The subsistence and part-time farmers, although making up a considerable fraction of American rural society, are marginal in an economic and production sense. Half of the nation's farmers produce almost 90 per cent of the commercially sold food and fiber. While Extension in the United States is available to all farm families, the larger share of its attention, especially on the agricultural side, is given to these commercial farmers.

What Is Extension in the United States? In a sentence, Extension *is* what Extension *does*. It can only be this because it is based on the needs of rural people as understood in the states and counties, and these needs reflect the varied conditions of American rural life and culture already noted. Extension, as its present federal director, M. L. Wilson, says, means "better homes and better farms to feed, clothe, and strengthen the Nation." It has also meant better organized, better functioning communities. It has proved that when the force of education is released in homes and on farms, in communities and counties, social processes are affected and social change takes place.

Put another way, extension is simply the effort to put the vital information of the agricultural scientist, whether he be physical or social scientist, at the disposal of the farmer and his wife. It is the battle to narrow the gap between ever-advancing knowledge and practice, whether the content of the knowledge relates to soil analysis or conservation, to animal husbandry or human nutrition, to community organization or parent education. "Extension" is simply a word used to indicate the whole complex of activities which enter into a program that is educational in its philosophy, its focus, its objectives, and its methods. It is preeminently a method and a process, not a system. It is a program that has developed over thirty years and always it has been built upon the needs of rural people. Its strengths and its faults in considerable measure grow out of this fact. It is never far from the grass roots—the farm people.

Objectives. Stated in brief form, some concrete objectives of agricultural, home economics, and 4-H Extension in the United States are:

1. To bring the farmer the knowledge and help that will enable him to farm still more efficiently and to increase his income.

2. To encourage the farmer to grow his own food, set a good table, and live well.

3. To help the members of the farm family to a larger appreciation of the opportunities, the beauties, and the privileges of country life, and to know something about the world in which they live.

4. To train youth to take his place as a member of the family, community, and society.

5. To promote the social, the cultural, the recreational, the intellectual, and the spiritual life of rural people.

6. To place opportunity before rural people whereby they may develop all their native talents through work, recreation, social life, and leadership.

7. To build a rural citizenry, proud of its occupation, independent in its thinking, constructive in its outlook, capable, efficient, self-reliant, with a love of home and country in its heart.

HISTORY AND FINANCE. Extension work in the United States is known as "Cooperative" Agricultural Extension Work because the Federal Government cooperates with the states, and through them, with the counties in its conduct. Extension work in the United States began in a nationwide way in 1914, following the enactment of a Federal law known technically as The Cooperative Agricultural Extension Act and popularly as the Smith-Lever Act. The law grew out of the experiences of the State Agricultural Colleges and State Departments of Agriculture in popular education outside of the schoolroom over a span of fifty or more years in farmers' meetings, lectures, and Farmers' Institutes; and about ten years of farm demonstration work by the United States Department of Agriculture, working at first largely in the Southern states and gradually spreading into other sections of the country.

Extension started in a relatively small way. In the second year the Federal appropriations, which had to be matched by the states, barely exceeded a million dollars. At the end of ten years of work the budget had grown to $8,680,000, just over half from Federal sources. For the fiscal year 1944–1945 the cost of Extension was approximately $37,000,000, with about $19,000,000 coming from Federal sources, $8,500,000 from state and college sources, $8,200,000 from counties, and a little over a million dollars from farmers' organizations. In addition to the above, the Federal Extension Service in Washington had nearly $750,000 for administering the Cooperative Extension Service and coordinating its Extension material with that of the states.

ORGANIZATION IN THE UNITED STATES. The strength of Agricultural Extension in the United States lies primarily in its county extension agents—agents who have their office in the counties and live among the people they serve. These agents consist usually of a county agricultural extension agent to work primarily with farmers and rural youth; a home economics extension agent, generally known as a home demonstration agent, to work with rural women and girls on all matters affecting the home; and sometimes a county agent who works almost wholly with rural boys and girls ten to twenty years of age in improving farming, homemaking and rural life. The agents who work with youth are generally known as 4-H Club agents or assistant county agents. Agents are usually farm reared graduates from a college of agriculture or home economics and have a practical knowledge of farming or homemaking and a liking for country life.

The county agents are aided and supervised in their work by supervisory agents employed by and located at the State Agricultural College. They are given help in their technical problems in the county by a corps of extension specialists, also employed by and located at the State Agricultural College, who are available on call by the county agents and who make visits from time to time to the counties to bring to the county agents the latest results and most approved methods of the college and its experiment stations for promoting improvements in agronomy, horticulture, forestry, dairying, animal husbandry, nutrition, farm and home management, and other like subjects.

Administering all extension work in every state is an Extension Director selected by the State Agricultural College and acceptable to the Federal Government. The Federal Government has a corps of technical specialists who aid the states in their plans for extension work and carry to the states the findings of the research departments of the Federal Government and keep the states informed as to the studies and observations by Federal extension agents on extension organization methods, policies and results throughout the entire United States.

At the head of the Federal Extension Service is an Extension Director, who administers the Federal staff and to whom all state plans for extension work, in which state funds enter into cooperation with Federal funds, are submitted for consideration and approval.

All extension agents, whether specialists or county agents, so plan their programs as to teach both adults and juniors.

A vital part of the county extension agent's plans is the voluntary enlistment of successful farmers, farm women, rural teachers, rural ministers, businessmen, and others interested in rural work as aids in helping carry the message and the better way of farming, homemaking, and community life to their neighbors. The county agents and extension specialists give special help and training to these men and women, usually known as "local leaders," who serve without pay and who constitute the backbone of the Extension Service in the counties.

As of July 1, 1944, there were 3,971 county agricultural agents and assistants, 2,579 home demonstration agents and assistants, 332 county 4-H (rural youth) club agents and 1,645 specialists. The remainder of the total staff of 9,180 were engaged in administrative and supervisory responsibilities. These figures do not include war emergency farm labor and other assistants.

In addition to this paid staff, there were likewise in 1943 over 1,066,000 leading men, rural women, and older youth serving as unpaid local leaders in assisting extension workers in carrying out extension teachings. Some of these people, in addition to several hundred thousand others, were also serving as neighborhood leaders to keep their neighbors informed on war programs. Valued at no more than farm labor wages, the voluntary contribution of these leaders, aggregating well over twelve million man-days a year, is worth more than the total cost of the Extension Service to all the contributing levels of government combined.

It is apparent from the above that extension is closely tied in to local situations wherever it functions. This in itself is a reflection of the cultural organization in rural America. The farmer, each on a separate homestead, is one to whom local concerns are important. He is conscious of the differences in his soil from his neighbor's or from that in the next county. He experiences world socioeconomic trends in his own community and as they affect his own farm operation and family living. He and his community must make their own adaptations to them. The type of organization described makes it possible for extension to channel information down to the individual farm and also for the problems of individual farm and local community to receive the attention of professional rural adult educators and county agents, and through them to come to the attention of specialists and research workers in the state, and if necessary, national, offices.

FARMERS' NEEDS AND EXTENSION METHODS. It has been stated that the normal program of extension has been, and is, the meeting of the needs farm people have, helping them to solve their problems. Shrewd common sense, plus trial and error in the early years, produced a pattern of methods and organization that enabled this to be done in harmony with the interests and values of the farm population.

The intense and natural interest of the farm family in its own land and home was met by emphasizing the importance of home visits by the agents. These visits were educational in character. The practicability of a suggested new practice was discussed in terms of the specific situation. With the passing of the years the objectives and scope of the home visit have necessarily broadened, but the device is unlikely to be discarded. It fits the value the American rural dweller places on neighborliness, on face-to-face contact, on the unique worth of the individual as such.

The farmer is a man who must constantly act. There are always countless things to be done on a farm and in a home. They are therefore most easily interested, especially in early contacts with extension, in information which they can immediately put to use in their farm or home practice. One of the earliest methods was designed to take advantage of this attitude. It was that of demonstration. These demonstrations were of two kinds, those dealing with methods and those dealing with results. In the former, a demonstration is given by an extension worker or other trained leader for the purpose of showing how to carry out a specific practice; its primary purpose is the teaching of skills. These demonstrations are attended by those interested. They may be specially set up or form part of the program of an adult or 4-H club.

A "result demonstration" may be illustrated by demonstrating that the application of fertilizer to cotton will result in larger and more profitable yields, that underweight of certain children can be corrected through drinking one quart of milk per day, that the use of certified seed in growing potatoes increases yield and quality, or that a larger farm business results from a more efficient use of labor and machinery. The chief purpose of the result demonstration is to establish confidence in the recommended practice. It is conducted by a farmer, homemaker, boy or girl under the direct supervision of the extension worker, to show locally the value of a recommended practice. Such a demonstration involves a substantial period of time and records of results and comparison, and is designed to teach others in addition to the person conducting the demonstration. This method, a unique feature of extension in the United States, is in effect learning by doing.

This teaching, whether by demonstration or through the more conventional methods of lectures at meetings, publications, news articles of the new devices of radio or moving picture, is based, as already indicated, upon research. Had we not had one or more agricultural colleges in every state in the Union to train men and women in agriculture and home economics and teaching methods, had we not had one or more agricultural experiment stations associated with these agricultural colleges in every state and territory in the United States, and had we not had a great Federal Department of Agriculture to do research work in every field of agriculture and home economics, extension could never have made the progress it has made. Research gives it the basic knowledge it extends. Extension is also based on local knowledge and experience. Extension in practice studies as well as extends. The county agents constantly survey, analyze, evalute the results of suggested new procedures and practices, and make their findings available not merely to their local constituents but also to the experiment stations at the agricultural colleges.

The development of the program shows clearly how it responds to recognized needs and changing conditions. Soon after its inception as a national agency, the nation was engulfed in the first World War. Naturally there was major emphasis on the techniques of increasing agricultural production and on the efficient handling of food and fiber within the farm home. The curriculum was largely vocational. After this conflict there was a serious agricultural depression. Farmers became interested in the possibilities of the cooperative movement as a means for improving their economic position. Extension began to give information and teach about cooperation and in a relatively short time the number of cooperative associations in the United States doubled. Extension was clearly an important factor, although not the only one, in this development. The intensification of the depression in the 1930s produced a number of new Federal agencies and a new national agricultural policy. Extension not only cooperated with the new agencies, but was heavily involved in explaining the new program to the farm population.

It responded to other demands. The interests of farm people broadened. Women became interested not only in the efficient preparation of food but also in the newer knowledge of nutrition; not only in making clothes for their children but in their children as human beings as well. From this it was a short step to questions about health and to moves for enriching the social life of farm people. Extension staffs, Federal and state, began to add specialists in parent education, in recreation, drama and music, and in older rural youth programs. Farm people became convinced that many of the problems they faced lay outside the line fences of their farms and were related to national and international conditions, trends and policies, about which they should be better informed. Tens of thousands of voluntary discussion groups were formed under extension auspices, supplied with impartial materials on controversial social and economic issues and launched on a program of considering public affairs. It was soon seen, as issues were clarified, that these new trends affected local situations, and farm people began to ask for assistance in local planning, sometimes only in terms of land utilization but quite frequently in terms of local institutions, social as well as economic.

The core of the program is still in terms of what may be called the practical side of agriculture and home economics, but the brief sketch of its development indicates that extension is sensitive to and responds to not only the needs of farm people but to the changing climate of public opinion in rural America. Indeed there is constant interaction between these, for extension teaching and leadership becomes an important factor in social change because it modifies the attitudes and beliefs of farm people by the steady stream of information its educational work brings to its constituents.

It is sociologically and culturally significant that extension has been called upon by the farm people to take on these newer aspects of its program. Other agencies could have been formed to undertake this work. But there are fewer agencies either available or possible in rural society than in urban. The Extension Service had won the confidence of its constituents to a considerable degree. The farm people turned to it because their prior experience with it seemed to warrant the belief it could serve the newer, changing needs and because it was already established and functioning.

These newer emphases have forced some adaptations of old teaching methods, the development of new devices, and the greater use of some of the more conventional teaching techniques. One cannot, for instance, have a result or method demonstration of the pros and cons of the issue of high tariffs versus free trade. Talks, forums, and discussion groups have come into prominence.

Comparably, there has been a decided trend in extension work away from individual farm and home service and visits to increasing work with organized groups of people in meetings and clubs. These groups may be concerned with special interests or they may deal with the total community. This trend has developed in part because of the greater mobility and travel range of farm people, made possible by the automobile and improved roads, in part because of the great increase in the number of volunteer leaders and their greater service to their neighbors, in part because of the very success of the extension program in the past, and in no small measure because of the fact that many of the newer features of the program cannot be economically or even effectively handled by the older, more individualized methods, and in part because of advances in the science of social organization. The extension worker, therefore, needs to apply the techniques of social or community organization. There is here an interesting case study of a large social institution like the Extension Service making adjustments to changing social and economic conditions affecting its field of service.

Some illustrations may make this clearer. In a certain disadvantaged rural community of 850 small farms, poor health of the women could be traced in large part to back-breaking drudgery, especially the washing of clothes, which was usually carried on by the spring or stream. This involved heating water by large bonfires, which, in turn, necessitated hauling a cartload of wood each wash day, and also created a fire hazard, caused a poor working posture and unpleasant working conditions because of the smoke. The people did not realize the need for improving the methods; they were simply following a time-honored routine.

The extension agents had won the confidence of these people, who had already organized fifteen home demonstration clubs and thirteen neighborhood committees. The problem, when brought before these groups, was recognized and studied. State specialists in home management and agricultural engineering were called in and met with the local people. An over-all solution was presented—the building of wash-pot heaters. This process took five months. The solution was presented at neighborhood and home demonstration club meetings and by local leaders and the agents in their home visits. Working drawings for the wash-pot heaters were given only on request and these requests were followed up by visits. Many persons requested drawings and built the heaters.

After using the wash-pot heater for heating water on hog killing day, the husband of one club woman said to her, "You know, this was the easiest hog killing day we have ever had!" She replied, "Yes, and just think how much work and wood we would have saved if we had built this wash-pot heater when we were first married!" One specialist concerned reported, "The pride and pleasure of accomplishment when families completed building the wash-pot heater proves its value. Enthusiasm for the 'drudgery-saver' is spreading as rapidly as one is completed. It is surprising how easily low-income families can use old brick and stone to build this convenience."

In a New England township twenty people met to plan for a better rural life. It wasn't long before they had discovered their two most pressing problems. First, the fact that the river had silted in and was holding back the run-off water, destroying about 900 acres of the most valuable crop land in the township. This meant the abandonment of farms or unprofitable farming, impossibility of collecting taxes, and the deterioration of the township in general. The township committee explored ways of solving this problem. First, they asked help from the Soil Conservation Service. Since the law under which this agency operated would not permit them to do the necessary job, a large chemical company was asked to blast a ditch through the area. The Company agreed to put on a demonstration if the farmers would pay for the materials used. The township committee then appointed a sub-committee to visit each farm owner affected, to have him sign an agreement to contribute at the rate of $3 per crop acre drained on his farm. Some blasting was done, but about this time the state passed an enabling act making it possible to set up Soil Conservation Districts. The township became part of a district and the Soil Conservation Service took over the task of dredging a ditch two and one-half miles long through the area. About 336 acres of crop land benefited directly and 236 acres indirectly from the drainage project.

The second problem was the expense of maintaining roads, schools, and other public services to residents in the poorer farming sections. Many of the properties in these areas were tax delinquent. The township Agricultural Planning Committee organized a program whereby the town would take over the ownership of these properties and establish town forests on them. The committee presented their organized program to the annual township meeting and secured the necessary legal clearance for the township to carry out the program. By special arrangement the United States Department of Agriculture, through the agricultural conservation program, paid the township a nominal sum for setting out the trees. Most of the farms taken over had been owned by the residents. These people have left these submarginal areas and are now employed in industry or are on farms in the valleys. At no expense to the township they are now making a much greater contribution to our national welfare and are maintaining a standard of living much higher than was possible on the submarginal farms.

The two last examples indicate how one township, through an organization sponsored by an educational institution in the state, outlined and overcame their most serious problems—problems which might possibly have meant bankruptcy to the township as well as to the individuals involved. The first example shows the effectiveness of extension in meeting a more traditional problem. By such methods, both in agriculture and home economics, millions of practices on individual farms and in farm homes are improved and changed every year. The total impact over the years is a significant degree of social change.

The use of hybrid seed corn is an example on state and regional bases. In 1933 only one tenth of one per cent of the total corn acreage in the United States was estimated to be of hybrid varieties. By 1943 this had been raised to 51.6 per cent and in some states, like Iowa, approached the 100 per cent mark. In this increase and diffusion through the corn belt, more agencies than extension played a part, although extension made it a high-priority project, using many devices to influence corn farmers to try the new seed. An Iowa study [1] shows that salesmen and radio advertising gave the original knowledge of these new varieties to three farmers out of five. But the most influential cause of adopting this type of seed was the success of neighbors. This influenced 45.5 per cent. Another 6.6 per cent adopted this seed after personal experimentation. More than half, therefore, were influenced by a demonstration, whether actually under extension auspices or not.

AGRICULTURAL EXTENSION IS DESIGNED TO BUILD RURAL PEOPLE AND TO IMPROVE AGRICULTURE. Extension is an educational agency that not only helps rural people to increase their efficiency and their income but also helps to build these people themselves into understanding, accomplishing, self-confident, capable men, women and youth, with vision and leadership. This building of rural people is the ultimate purpose of Extension.

Agricultural Extension in the United States makes rural people partners with government in the selection, financing, and direction of local extension agents. This is an educational process. Farmers and their families take part in making local surveys for getting and interpreting local practices and data upon which local and state extension

[1] Bruce Ryan and N. C. Gross, "Journal Paper J-1092, Iowa Agricultural Experiment Station," *Rural Sociology*, March, 1943.

programs are integrated and built. This is an educational experience. The farmer carries out the demonstration and explains its meaning to his neighbor. He helps in the development of rural organizations that may best serve his needs and those of his family. He is encouraged to take part in committee work, to speak at meetings, hold office and do other work that will give him experience in guiding a cooperative or promoting the social, recreational and civic welfare of the community. These are educational processes that help to bring satisfaction to tne individual and give meaning to rural life. Extension that does not have for its ultimate purpose the building and growth of rural men, women, and youth has not caught the spirit of extension but is dealing with its bones.

Chapter 14 · THE ROLE OF EXTENSION IN WORLD RECONSTRUCTION · By *M. L. Wilson* and *Edmund deS. Brunner*

THE PRECEDING CHAPTERS have shown that rural adult education with special reference to agriculture and somewhat less emphasis on home economics is a world-wide phenomenon. While it is usually a governmental activity, at times various types of private agencies engage in such work, particularly in countries not yet under the sway of modern scientific developments. Many such countries are also economically disadvantaged. The name frequently applied to such rural adult education is Extension.

It is clear that the objective of the educational work of these extension services is to produce better homes and better farms to feed, clothe, and strengthen the entire population of an area or a nation. As extension work develops and wins the confidence of its constituency, it also becomes interested in and helps to produce better organized, better functioning communities. This book is replete with illustrations of the fact that, when education begins to operate in farms and in homes and communities, practices change. Such changes influence and interact with other aspects of the society and its economy. Social processes are thereby affected. The educative results are cumulative, and significant social change occurs. Thus extension teaching at its best, regardless of its administrative arrangements, is a social process, not a system or a program to be administered. It is everywhere different but everywhere similar in the principles of its operation. The differences are those which grow out of variations in environment and culture. The similarities arise from the common needs and experiences of human beings.

This contribution of our extension service through its teaching is made by applying to the problems of the people on the land scientific knowledge derived from research. Science is based upon the assumption of a uniformity in natural laws, and certain predictions can be made in accordance with those laws. It is, therefore, wholly impersonal and knows no geography nor group of people. If scientific knowledge is directed and controlled by certain philosophic and religious views with reference to the ultimate purposes and values in life, we believe it to be a great blessing to mankind. On the other hand, if proper ethical goals and values do not control science, it can be a terrifically powerful instrument for destruction and human misery. From the point of view of the cultural anthropologist, all human cultures have a scientific sector. No matter how primitive a culture may be, as judged by our standards, man has always sought to exercise in some manner some control over his environment, his needs, and desires.

Scientific knowledge has, with the past one hundred and fifty years or so, expanded with incredible speed. Man at the moment is in the process of great cultural change, whereby scientific knowledge and its application are pushing out prescientific folklore and the folkways that go with it. In the field of agriculture, where man exerts control over the forces of nature and directs them in the production of food and fiber, advances have been as rapid as in any field.

Extension, in common with all other organized social effort now faces a future never more laden with an unparalleled number of serious problems, many of them unpredictable. Shells and bombs have destroyed thousands of coconut palms and breadfruit trees on remote Pacific islands. Armies and their machines have scarred the well-tilled fields of Europe and China. Populations have been uprooted. The farmers' plowshares have been beaten into swords. Some populations have been driven to the brink of starvation. First relief, then rehabilitation loom as major tasks in the period just ahead. How can they be accomplished without the cooperation of those peoples less scarred by war, and especially without the cooperation of their farmers? In this situation Extension as a rural adult educational enterprise has an important contribution to make. In fact it is the conviction of the authors of this book that since extension work has been tried successfully in so many parts of the world thus far, it can be valuable to a high degree wherever its underlying principles are applied. It is the purpose of this chapter to summarize the guiding principles of extension that appear from the foregoing chapters to have wide applicability to the problems which those who serve the rural peoples of the world will face in the reconstruction period.

The basic principle in extension work when approached as a stimulating process is that any program must be in harmony with the culture of the people with which it is concerned.

Culture involves the whole of man's social behavior as it can be observed in the group life of human beings. This includes 1) the machines and skills and methods by which men make a living, 2) the habits and organizations which people must accept if they are to live together in a community, and 3) religious beliefs, values, and practices. In short, the culture consists of the habitual and systematized ways by which people make a living, of the social organizations through which they cooperate to achieve their mutual desires, and of their attitudes, values and faiths, their sanctioned ways of life. No people are so primitive that they lack a culture. Thus the aborigines of Australia, one of the most primitive of peoples, have one of the most intricate kinship patterns.

The culture in its totality is very precious to any community or people because their accustomed behaviors are rationalized in terms of the values held. The simpler the society, the less experienced it is in meeting the impact of change, the more tenaciously will it cling to its culture. It is the one thing its people can keep measurably intact, so they believe, in their struggle for survival and freedom. Their sense of security is bound up in it. The cooperation of the extension leader or teacher and the people cannot be continuously effective unless the culture is understood and the work carried on in harmony with it. Foolish as some of its mores and taboos may appear to the scientifically trained outsider, each culture has behind it centuries of effective operation in terms of the environment of those who own it. It has developed out of their past experiences. Its sanctions are rooted in hallowed history. They have no basis for envisaging the new and the strange. The impact of the new age of science, its ultimate promise, once the effects of dislodging the old patterns are absorbed in the ongoing processes of a society, must be revealed with sympathy and understanding.

It follows that the next basic principle of extension work when it is oriented to the culture and recognizes the existence of folk knowledge is that of democratic operation. A dictated program which goes against the cultural values of a community or a whole society meets a resistance based on this very culture. If the program is backed with sufficient power, it may be carried out. If that power is removed, the people will revert to their former ways. Only if power is applied for years can a superimposed policy succeed. Democratic operation is usually the speedier way to achieve the long-time goal if the objective is in terms of individual and community well being and not in terms of a program as such. Moreover, the democratic process builds good will rather than breeding discontent or resistance. It is usually not difficult to rationalize a desirable program in terms of a local culture and its "pillars of policy." Through extension education old values may give way to new and create desire for and satisfaction in change.

In the emergency situations that may in the immediate future confront extension workers, relief workers, and especially health workers, it may at times be necessary to exert pressure to reach an objective. In this case the pressure should be applied by an administrative agency and not an educational one. The greatest possible effort that the circumstances permit must be put into continuous education as to the desirability of the program. If pressure is used, it is highly desirable that its objectives be tied to a long-range program supported by education. For there is evidence aplenty in this book that when the people realize that the service of extension is theirs—for them and their good—and that when they share to a maximum degree in the planning and operation of programs, success will be rapid, continuing, and pronounced.

This principle of democratic operation recognizes that extension must be a give and take process. It recognizes that the accepted ways of communication are a part of any culture and that these ways must be used and must work as a two-way process. Local people always know more about some things in the environment, more about the reasons for some problems than any worker from the outside. For instance, in one Pacific island, taro, which has a role somewhat like that of the potato in Euro-American agriculture, seems to exhaust the soil in from one quarter to one half the time it takes for a similar process even in neighboring islands. Obviously this affects the whole scheme of rotation. Conceivably the worker from the outside might not know this; the local people do. This local knowledge if sought after and received, not only facilitates the total program, it also often furnishes important research problems for the agricultural experiment stations, the results of which can be fed back through extension to the local people to improve further their agriculture and their life.

Implicit in the principle of democratic operation is the additional principle that extension must work with communities and all people within them. This holds true not only because national and international social and economic forces are experienced by most people only where they live; it is important also because communities in the nature of the case are associations of people. A weed or a pest on one farm spreads to others. Products of poor quality pull down the standard of a community bulk shipment to market. For reasons such as these, extension must work with the large landholder as well as the small, with the tenant operator as well as the owner, with the family on the back road as well as the one on the main highway. Extension cannot be class conscious and wholly successful at one and the same time. It cannot work for one group and not all groups.

Moreover, there are many needs that are community-wide. True, the peculiar problems of individual farms and homes must be kept in mind, but as farm families become more and more familiar with the influence of world and domestic forces and policies, as they learn more about soil conservation, as they increasingly plan the kind of life they want for themselves and their children, they will expect the community increasingly to become the focus of many of the programs.

But while this is true, it is equally true that extension must work with families. The family is central in folk societies and it is still highly important even in Euro-American society. Farming is a family enterprise in which every able-bodied member has his share. The family should be the unit for extension work. In highly developed extension services what the specialist proposes for one aspect of the farming operations may not only upset the balance of the farming operations, it may be bad for the family as a unit. There are times when buying a washing machine and a motor dishwasher may be better from the family point of view, even in economic terms, than purchasing a new tractor.

It is a cardinal principle of extension that the program must be at once simple in conception and comprehensive in scope. It must be built on the vital, recognized needs of the local people concerned. It is of course the part of extension leadership to help the people to recognize needs which they do not appreciate. The iron grip of tradition and custom can often be broken by skillful teaching. But especially at the inception of extension work it is important to start with projects the importance of which is either well recognized or can be easily demonstrated. If the initial projects are simple and easily successful, the people gain confidence both in the professional leader and in the organization which he represents. They are ready then to go on to more difficult and often more important projects. Thus the program evolves and progresses, especially where the skillful leader makes wise use of the local situation and of materials ready to hand and familiar to the people. The value of basing programs on something—

or relating them to something—that the people already have or know about cannot be overestimated.

The program, of course, should not be built by the professional leader alone. Participation of all concerned in program planning is of prime importance in extension, since so much of learning takes place by doing. It follows that cooperation of professional worker and laymen is equally important. The former cannot reach personally every farm home in his district frequently enough to do effective teaching. Therefore if he can train leaders to help carry the program, he multiplies his efforts manyfold. These leaders of course benefit from the training they receive, but they are able to pass on the experience. It is well known that the testimony of one local person to another about the desirability and practicality of a suggested procedure ranks very high on the list of reasons farm people give for changing their practices. The problem of selecting, training, and supervising these volunteer leaders cannot be entered into here but it should be said to the skeptical that leaders can be found. There is no culture in which the function of leadership is not present and exercised.

It is also important, on some subjects more than others, that extension work should be carried on with groups on a community or neighborhood basis. In the areas of the world where good roads and automobiles are fairly common this is relatively easy and a considerable timesaver. But it is not too difficult elsewhere if rural society centers, as it so largely does, in the village from which the farmer goes out to his fields. Village meetings have been developed to a very high degree, for instance, in Japan.

Two other principles should be mentioned in relation to community programs and the whole area of program building. The first is that the simpler the society, the broader the general program may need to be. In a simple society there will be few social agencies but there may well be many needs. Perhaps the first *economic* need of a given population would be to rid it of hookworm infestation and to persuade the people to wear sandals to prevent reinfection. When this was accomplished on a rubber plantation in Ceylon some years ago, per worker production and wages went up and the plantation costs per pound of latex also went down. Thus health, nutrition, education and many other things enter frequently into a well planned extension program. Needless to say, if any existing agency is charged with meeting such needs, the function of extension is then simply to enlist the interest of that agency. Indeed, another guidepost is that extension service should co-operate with other agencies, whether secular or religious, engaged in rural service. Where such agencies are local and precede the coming of the extension service, it is far better to use them than to try to build a new one.

The illustration indicates another principle. Local people think in terms of themselves and their communities. The well-balanced pro-gram, therefore, should be as broad as the needs of rural life, but there is more than enough work for all agencies, if the extension service is fortunate enough to have others to which it may turn. In either case, whether the work is done by extension alone or in cooperation with others, as Dr. James Yen says: "The problems must be correlated, the programs integrated."

Another guidepost in extension planning is that the programs adopted should fit the budget or the budget should fit the program adopted. Many good extension workers and programs have come to grief even after an auspicious start because more than the impossible was attempted. The local people soon become discouraged under such conditions. In this connection, it should be noted that there is much of value in having the beneficiaries of the program share in its support, even though to a relatively minor degree. It thus becomes more truly democratic and cooperative.

This discussion has not thus far dealt with methods and techniques of extension education. These will vary according to circumstances. They are all conditioned by whatever laws govern adult learning. But there is one method so often used, so often praised the world over that it has become an educational principle of extension everywhere—the demonstration. This has been defined and illustrated in earlier chap-ters. Here it need but be stated that extension dares to put its teachings to the test, not only in the laboratory but in the practical situations on farms and in homes. Its primary job is to put the knowledge of the scientist, physical and social, at the disposal of the farmer, his wife, and their community in such workable terms and by such tested principles that a life better than was known before will result. This is extension's contribution to the task of reconstruction.

INDEX

Aborigines, absorption or extermination of, 9

Adult education, rural: extension applied to, 193, 194; *see also* Agricultural education

Advisory service, *see* Agricultural extension

Africa, problems of British in, 18; impact of European civilization on tribes, 10

African National Congress, 12

Age, Arab reverence for, 82

Agrarian legislation, 146, 148

Agricultural Advisory Service of United Kingdom, 155

Agricultural colleges, in the United Kingdom, 154; scholarships, 156; and experiment stations in U.S., 186

Agricultural education in northwest Europe, 169-74; primary, 169; Winter Schools, 170, 171; middle schools: colleges and university departments, 171 ff.; why institutions have not produced results hoped for, 172; how to improve effectiveness, 173

—— in United Kingdom, 154

Agricultural expert, titles used in different countries, 2

Agricultural extension, what extension is, 1-7, 191; promoted by governmental authority, 3; how sound practices are extended, 3; basic objective, 5; effective work relies on local leadership, 55, 71, 96, 113, 136; local material should be used, 56; importance of choice of site, 67, 133; and of attitude of people, 67; people suspicious of governmental programs, 73; work should be integrated into life and culture of community, 132, 134, 135, 196, 198; plans for postwar unification of all branches in United Kingdom, 162; national financing, 163, 176; must be based on self-help, 177; number of rural, farm, and other families influenced by, in U.S., 180; closely tied to local situations, 185; role in world reconstruction, 193-99; family the unit for, 197; should cooperate with other rural service agencies, 199; *see also* Demonstration; Leaders; Programs; *and under names of countries*

Agricultural extension education, based upon cultural variation and change, 1, 4; a rural adult educational enterprise, 193, 194

—— workers: titles vary around the world, 2; usually civil servant, 3; rural background needed, 73; ideal goal, 96; in the Balkans, 109, 114-16; preservice and in-service training, 115; augmented to meet wartime needs in United Kingdom, 159; selection of, 163, 178; techniques of social or community organization, 189; *see also* Leaders

Agricultural Organiser in United Kingdom, 155, 157

Agricultural press in United Kingdom, 161

Agricultural societies, annual journals, 161

Agriculturists, regional, in Latin America, 127

Albania, population pressure, 101

Allen, H. B., 112

Allotment associations, 161

American Republics, *see* Latin America

American University of Beirut, Institute of Rural Life, extension work, 86, 93

Andean countries, extension work among Indians of, 129, 133; colonization, 130; health and sanitation, 132

Anglo-American culture, *see* Euro-American society

Animal husbandry, effect of introduction among Navajos, 16

Apache Indians, 15

Arab *fellahin*, extension work among, 78-100; emotional attachment to land, 79; cultivate land as tenants or share croppers, 80; religion, 81; main culture values, 82; illiteracy, 83, 85; economic condition, 84; emotional appeal most effective technique, 91; influence of Koran: concept of honor: past, 92; spirit of nationalism, 93

Arab village community, 79-83; physical structure: pillars of its culture, 79; family, 80, 92; religion, 81; need for extension, 83-87; need for more schools: health situation, 84; limited extension experience, 84 ff.; in Palestine, 85; in Lebanon, Syria, Iraq, 86; suggestions

Arab village community (*Continued*)
for effective extension, 87-97; initial project, 87; choice of center, 88; local situation, 89; projects, 90; special techniques, 91; utilization of existing organization, 95; of local leadership, 96; village pump, 97-100

Argentina, Indian population, 118; independence of peasants, 120; specialists: extension and the schools, 128; extension and colonization, 130

Artisans' contribution to welfare of folk society, 39

Australia, agricultural extension, 5; aborigines have intricate kinship patterns, 195

Ayllus, group solidarity, 121

Bailey, L. H., 48

Balkans, extension in the, 101-16; economic background, 101-4; population pressure, 101; depletion of soil: subsistence agriculture: commissation, 102; regulations of emigration: development of industry, 103; social background, 104; cooperatives, 105; research and education, 105-7; existing extension systems, 107-10; traveling conditions, 109; essentials of better extension, 110-13; pivotal points for reorganization, 111; organization of youth, 112; personnel, 114-16

Banks develop experiment station and extension programs, 129

Bantu, impact of European civilization, 10; churches: trades union, 12

Barter economy of folk society, 41

Batavia, overpopulation, 33

Bedouins, 78

Beekeeping in India, 70

Beirut, extension work of American University's Institute of Rural Life, 86, 93

Belgium, agricultural extension, 1; illiteracy, 167; agricultural education, 169, 172; Winter Schools, 170, 171; middle schools: home-science schools for girls, 171; colleges and university departments, 171; provision for service, 174; *Boerenbond* advisory service, 176

Bolivia, Indian population, 118

Bookkeeping, farm, 175

Boys, farm projects for, 3; 4-H clubs, 3, 130, 183, 184

Brand, Donald, 117

Brazil, Indian population, 118; independence of people, 120; Farmers Weeks: specialists and campaigns, 128

Broadcasting, 94; as adjunct to education, 161

Bucher, Karl, 28

Bulgaria, population pressure, 101; cooperatives, 105; poultry raising: specialists in plant protection and homemaking, 113

Canning, 176

Cattle in India, 64

Ceylon, effect of eradication of hookworm, 198

Chile, Indian population, 118; independence of peasants, 120; extension and the schools, 128; work of church missions, 130; home economics, 130

China, rural reconstruction, 1; agricultural extension, 46-60; nature and importance of agriculture: rate of productivity, 46, 47; size of land holdings, 47; cultural stability, 47 ff.; Mass Education Movement, 50; contribution of universities and colleges, 50; Governmental agencies, 51; programs to date, 52-55; buying foreign machines for exhibition, 54, 56; importance of local leadership, 55; using local materials, 56; programs wrecked by overstaffing, 58; cooperation with other countries indispensable, 60

Chinese, diet: social behavior, 47; attitude toward change, 48, 49, 57

Chinese Agricultural Association, outline for postwar agricultural reconstruction, 52

Church, influence upon folk society, 39, 40; importance of, in Spanish-speaking villages, 123

City, cleavage between village and, 42

Collectivism in Russia, 147

Colombia, Indian population, 118; independence of peasants, 120; extension work on campaign basis, and through schools, 128; work of church missions, 130

Community organization, in Arab village, 78-100

Community program, *see* Program

Cooperation, in Pacific islands, 30, 31; in Egypt, 88

Cooperative Agricultural Extension Act, 183

Cooperative Agricultural Extension Work in U.S., 183

Cooperative societies, in India, 66, 76; in northwest Europe, 166

Copenhagen, Royal Agricultural College, 172

Coral atolls, 23

Corn, use of hybrid seed, 191

Costa Rica, Indian population, 118; independence of peasants, 120

Cotton, standard, 143

County agricultural extension agents in U.S., 183; number of agents and assistants, 184

County Agricultural Service of United Kingdom, 155

Culture, effect of failure to appreciate differences, 8; defined, 9; beliefs and other traits interfere with extension programs, 68; involves whole of man's social behavior, 194

Culture change, 12, 194; extension education based on culture variation and, 1, 4; culture contact and, among nonliterates, 10-17; value of an innovation, 6

Dacus fly, campaign against, 113

Damascus, Village Welfare Movement, 94

Danube Valley Authority, 103

Demonstration method, 2, 186, 199; in India, 63, 66; in United Kingdom, 156

Denmark, farm ownership, 166; People's High Schools: high level of farming ability: cooperation and education, 169; home-science schools for girls, 171; Royal Agricultural College, 172; extension service, 174, 176; farm bookkeeping, 175; breed societies and cooperatives, 176

Depression of the 1930s, 187; effect upon Navajos, 17

Development Fund Act of 1909, 154

Deventer, Neth., college for tropical agriculture, 172

Discussion groups, in United Kingdom, 157; system of farm production records evolved, 160; in U.S., 187, 188

Diseases, extension activities help exclude, 23

Domestic Poultry Keepers Club, 162

Dutch East Indies, resettlement of Javanese in, 34

Ecuador, Indian population, 118; Indian Lay Health Program, 136

Education, responsibility to society, 6; *see also* Agricultural education; Agricultural extension education

—— in northwest Europe, 167-69, 172; establishment of rural high schools advocated, 168; Winter Schools, 170, 171

Egypt, population, 78; program to improve lot of peasant, 84; education: efforts for agricultural improvement, 85; cooperative movement, 88

El Pueblo, N.M., family rehabilitation, 123

El Salvador, Indian population, 118

Emergency situations, 196

England, *see under* United Kingdom

Environment, influence of geographical, 9, 178; specialized adjustment to a limited, 23

Euro-American society, phenomenal spread of culture, 8; two streams of influence, 14; rural society, 138-52; general traits, 138-45; produces goods primarily for market or for sale, 138; how individual differs from folk type, 138, 144 f.; term, 138*n*; farmer's view of land, 141; exceptions within culture, 145; cohesiveness, 147; regional variations: economic factors, 149; village or nonagriculturally employed person, 150; race as a differential, 151

Europe, regional variations in economic rationalization of farmer, 149; well-to-do farmers control most of best land and economic resources, 150; *see also under names of countries*

—— northwest: extension services, 165-79; land shortage: population density: size of farms, 166; educational system, 167-69; three aspects of agricultural extension, 167; rural schools, 168, 172; agricultural education, 169-74; how to improve effectiveness, 173; extension service provided by State or by farmers, 174-77; farm bookkeeping, 175; emphasis on democratic equality: influence of geographical environment: *see also* Belgium; Denmark; France; Germany; Netherlands; Switzerland

Extension, *see* Agricultural extension

Familistic societies, 39; *see also* Peasant societies

Family, central institution in most folk societies, 39, 40

—— rural, task of extension work to help, 1; in China, 48; in Arab village, 80; in the Balkans, 104; in Latin America, 121, 123; effect of decreased size of, on extension work, 144; the unit for extension work, 197

Farm bookkeeping, 175

Farmers, special agencies carry on educational activities among, 1; live by tradition, 4; humanity among, 5; challenge to organize, 5; Euro-American, 141; independent of clan or village group, 142; effect of size of family, 144; legislation applying specifically to, 146; cohesiveness, 147; regional variations in economic rationalization, 149; well-to-do, control most of best land and economic resources, 150; work among, 157; technical help for, in United Kingdom, 159 ff.; advice to, provided by organized activities, 175; skepticism and hostility of, 177; in the U.S.: standards of living: racial origins, 180; high mobility among: act independently of village or clan, 181; needs and extension methods, 185-91

Farmers' clubs, 157, 172

Farmers' societies, bookkeeping organized by, 175

Farming, "fire farming," 24; as a social system, 145n; collective, in Russia, 147; Germany's attempt to recreate a new form, 148; a family enterprise, 197

Farm institutes, in United Kingdom, 154; scholarships, 156

Farm projects for boys and girls, 3

Farms, number per county in United Kingdom, 155; typical size in northwest Europe, 166; number in U.S. and number of persons on them, 180

"Farm walks," 159

Farm women, importance of teaching better practices in homemaking to, 2; clubs, 158; associations of, organize fruit canning, 176

Fellahin, see Arab *fellahin*

Fertilizer, *see* Manure

Filipinos, 23; voluntary migration, 33

Films, as adjunct to education, 94, 161

"Fire farming," 24

Folk society, usually familistic, 39; Protestant influence on, 40; conservatism, 195; barter economy, 41; an integrated system, 44; characteristics of Western society, 45; Balkans, 104; moral code of, 114, 115; weights and measurements, 139; attitude toward land, 141; *see also* Arab *fellahin;* Peasant societies

4-H clubs, 3, 183, 184; in Latin America, 130

France, system of land tenure, 165; pattern of agricultural society, 166; elementary education, 167; agricultural education, 169, 170; Winter Schools for girls: middle schools devoted to viticulture, 171; colleges and university departments, 172; provision for financing, in *departements,* 172; extension service, 174; village *syndicat,* 176; farmer hostile to formal advice, 177

Fukien Christian University, Foochow, 51

Gamio, Manuel, 125

Gandhi, 75

Geographical factors, importance of, as related to retardation of agriculture, 125

Germany, attempt to recreate a new form of collective farming: probable effect of Allied occupation upon farmer, 148; system of land tenure, 165; pattern of agricultural society, 166; rural equalitarianism, 167; Winter Schools, 170; colleges and university departments, 171; extension service, 174, 176; farm bookkeeping, 175; home welfare, 176; farmer hostile to formal advice, 177

Gilbert islands, 23; growth of native cooperatives, 31

Girls, farm projects for, 3; Winter Schools for: home-science schools, 171

Great Britain, State system of extension service, 176; *see also* United Kingdom

Greece, population pressure, 101; tobacco, 113

Grignon, Fr., college for crop husbandry, 172

Group work, extension becomes education for, 6; in U.S., 188; community or neighborhood basis, 198

Guatemala, Indian population, 117

Hatch, Spencer, 133

Holland, college for tropical agriculture, 172

Home, *see* Family

Home economics extension agent, or home demonstration agent, in U.S., 2, 183, 184

Home projects for boys and girls, 3

Honduras, Indian population, 118

Honey, as medicine, 70

Hopi Indians, 9, 14

Hookworm, in India, 72, 198

Hungary, land tenure: population pressure, 101

India, rural reconstruction, 1; peasants follow economics of scarity, 44; extension experience in, 61-77; population, 61, 64; fragmentation or parceling of land, 61; methods of farming: characteristics of people, 62; need for rural reconstruction and extension, 63 ff.; village economy, 64; aid in research and experimentation, 65; making extension work effective, 66-76; selection of site, 67; formulating program, 68-71; beekeeping, 70; hookworm, 72; distinctions of caste and creed put aside, 75; problem of rural economy, 76; *see also* Martandam

Indians, Latin American: Hispanized, 9; programs for, 18; distribution of population, 117; conflicts re land tenure practices, 119; missionaries abolish and restore practice of whipping, 120; extension work of church missionaries, 129; in Andean countries, 129, 130, 132, 133; diseases, 134; many give up chewing coca leaves, 135; Indian Lay Health Program for Nicaragua and Ecuador, 136

—— North American, 181; effects of culture contact, 13-18

Inter-American Indian Institute, Indian Lay Health Program for Nicaragua and Ecuador, 136

International Missionary Council, 133

Iraq, population, 78; program to improve lot of peasant, 84; extension work, 86

Islam, protector of folk society, 40; a dominant factor in Arab culture, 81

Italy, agricultural extension: under Fascist regime, 1

James, Preston, quoted, 122

Japan, village meetings, 198

Japanese, in Brazil, 117; in U.S., 181

Java, 23; government pawnshops, 32; population increase, 34

Jibrail, Syria, 97

Junker, 150

Koran, influence of, 92

Kulak, 150

Labor systems, Pacific islands, 34

Land, folk society fertility rites, 38; emotional attachment of Arab *fellahin* for 79; principal types in Middle East, 79; tenure, 80

Latin America, extension work, 117-37; diversity of cultures, 117, 125; population, 117; cultural characteristics of work, 118-26; importance of personal relations, 118; peon-patron relationship, 119; literature, 119*n*, 125; lack of local responsibility for local and national welfare, 120; tax structure: importance of familism, 121; importance of village, 122; the church, 123; attitudes toward money: changes in agriculture spread slowly, 124; mountain and jungle peoples, 125; agricultural extension and research needed, 126; types of extension, 127-30; veterinarians and regional agriculturalists, 127; specialists and campaigns: extension and the schools: Mexican educational mission, 128; private agencies: demonstration farms and ranches: church missions, 129; extension and colonization: home economics, 130; looking ahead, 130-37; adjustments needed, 130-32; ten principles which should be followed, 133 ff.; *see also under names of countries*

Latrines, India, 72

Lawes, Sir John, 154

Leaders, utilization of local or volunteer, 55, 71, 96, 113, 136; advantage of farm background, 73; training of local, 184, 198; number in U.S., 185; understanding of local culture necessary, 195; *re* local recognition of needs, 197

Lebanon, population, 78; extension work, 86, 88, 90, 91

Legislation, agrarian, 146, 148

Li-Chuan Christian Rural Service Union, 50

Linguan University of, Canton Province, 50

Literature, Latin American, 119*n*, 125

Livestock-breeding societies, journals, 161

Malaysia, 20, 23, 24; rubber and coffee, 27; tenancy and usury, 32

Malinowski, Bronislaw, 9

Manila, overpopulation, 33

Manure, as fuel, 64, 88, 102

Market production a characteristic of Western society, 138

Markets, complicated price structure, 4

Martandam, India, Rural Demonstration Center, 66 ff.; poultry project, 68; purpose of work defined, 69; buildings, 74; Practical Training School of Rural Reconstruction, 74

Mass Education Movement in China, 50

Melanesians, 19

Mexicans indispensable to Southern agriculture, 181

Mexico, Indian population, 118; social revolution, 125; extension and the schools, 128, 129

Micronesians, 19

Middle East, political units, 78; no isolated farmsteads in, 79; land tenure, 80; religion, 81; invasion of Western culture, 83; limited extension experience, 84; youth eager for patriotic service, 87; educational campaign, 94 ff.; moving pictures: radio, 94; *see also* Arab *fellahin;* Arab village; Egypt; Iraq; Lebanon; Palestine; Syria; Transjordan

Milk testing, 160

Missionaries, work with Indians in Latin America, 120, 129; influence of Protestant, 135

Money, Latin American attitudes toward use of, 124

Money economy, 42

Monterde, Francisco, 125

Montpellier, Fr., college for viticulture, 172

Moving pictures, educational use of, 94, 161

National Farmers' Union, 157

National Federation of Young Farmers' Clubs, 157

National Milk Testing and Advisory Scheme, 160

National Peiping University, 51

National Southeastern University, China, 50

Navajo Indians, 15 ff.; effect of introduction of animal husbandry, 16; effect of drought and depression, 17

Needs, community, 197

Negro, Southern, 149, 151; indispensable to agriculture, 181

Netherlands, pattern of agricultural society, 166; rural schools, 167; agricultural and horticultural education, 169, 170, 171, 172; progress due to agricultural teachers, 170; Winter Schools: home-science schools for girls, 171; colleges and university departments, 172; extension service, 174; farm bookkeeping: cooperative management associations, 175; breed societies and cooperatives, 176

New Guinea, 19, 23, 24; depopulation, 33; Mandate, and indentured laborers, 35

New Mexico, rehabilitation and extension work among Spanish-speaking villagers, 123

Nicaragua, Indian Lay Health Program, 136

Niue, 33

Nonliterate peoples, oral tradition among, 4; culture contact and change, 8-18; regions occupied by, 9; extension programs, 17-18; *see also* Arab *fellahin;* Indians, American; Pacific islands

North American Mission Board, 120

Orientals indispensable to Western agriculture, 181

Pacific islands, extension work, 19-36; area, 19; population, 19, 21, 33; native economic systems, 22-25; isolated groups, 23; food, 23 ff.; "fire farming," 24; modern changes, 26-28; types of product supplied to world markets, 27; in years of depression and Japanese occupation, 28; persisting native organization and values, 28-32; social customs, 29; work habits: cooperation, 30, 31; lack of storage facilities: native handicrafts and industries, 31; special problems, 32-35; resettlement and colonization, 33; health and welfare conditions, 34, 35; labor systems, 34; future prospects, 35; destruction caused by bombs, 194

Palestine, extension activity, 84 ff.; Village Welfare Service, 86; cooperative movement, 88

Panama, Indian population, 118

Papago Indians, 15

Papua, 21

Paraguay, Indian population, 118

Paris, Institute Agronomique, 172

Parity, appeal of, to Euro-American rural dweller, 140; legislation necessary to meet demands for, 146

Peasant societies, characteristics, 4, 37-45, 124; spatial and mental isolation, 42; standard of living: fatalistic creed, 43; efforts at extension, 44; prestige of local leaders, 56; simplicity the keynote in efforts with, 73; intense individualism in northwest Europe, 178; *see also* Folk society

People's High Schools, Danish, 169

Peru, Indian population, 118; extension work carried on through schools, 128; medical and surgical clinic in Andes, 136

Pests, control of, 23

Philippines, 23; tenancy and usury, 32

Pima Indians, 15

Piute Indians, 14

Polynesians, 19

Pondo, impact of European civilization, 11

Poultry, project at Martandam, 68; community hatching and brooding of pure-bred chicks in Bulgaria, 113; industry in America, 140

Poultry clubs, 162

Primitive peoples, 8; absorption or extermination of aborigines, 9; slow to accept one of own people as merchant, 25; no people so primitive that they lack a culture, 195; *see also under* Nonliterate peoples; Indians; Pacific islands

Programs, and native peoples, 17-18; popularity depends upon local recognition of need for it, 69; long-time point of view: should build upon what people already have, 70, 95, 194, 195; selecting, 73; value of community projects, 91; wartime, in United Kingdom, 159; development, 187, 188; farmer and his family take part in, 191; must be simple and comprehensive, 197; importance of planning, 198; should fit budget, 199

Proletarization coincident with introduction of wage economy, 9, 12

Property holding, cooperative usages, 30

Provincial Institute of Mass Education, Fukien Provence, 55

Public health service in India, 66

Pueblo Indians, 14, 15

Puerto Rico, home economics, 130

Queensland, agricultural extension, 5

Race as a differential in Euro-American society, 151

Radio, 94, 161

Reconstruction period, role of extension in, 193-99

Redfield, Robert, quoted, 44

Rehabilitation a major task ahead, 194

Reisner, John H., 49

Relief a major task ahead, 194

Religion, place in familistic society, 39, 40

Rennes, Fr., college for dairying, 172

Research, 154, 187

Result demonstration, 186; *see also* Demonstration

Rich, willingness to copy the poor, 67

Roman Catholic Church, 123; has tended to preserve folk society, 39

Roman Catholic farmers associations, 171

Rothamsted Experimental Station, 154

Royal Agricultural College, Cirencester, 154

Royal Agricultural Society experimental station at Waborn, 154

Rumania, land reform: population pressure, 101

Rural reconstruction, 71, 76; in China, 50

Russell, Sir John, 65

Russia, mysticism and realism in attitude of serfs toward land, 38; collective farming, 147

Science, progress in agriculture and rural welfare, 7

Seed stores in India, 66

Share croppers, 141, 151, 180

Slavs, "Zadruga" system of, superseded, 104

Smith, C. B., 58, 127

Smith, T. Lynn, 117, 122, 126

Smith-Lever Act, 183

Society, transition from medieval to modern, 165; *see also* Euro-American society; Folk society; Peasant societies

Soil Conservation Districts, 190

Solomon Islands, 19

South Africa, impact of European civilization, 10; natives barred from owning land, 12

South Sea islands, 19

Spanish-Pueblo relationship, 14

Standards of living, in peasant societies, 43; among farmers in U.S., 180

State experimental farms, farmers skeptical of results on, 177

Sumatra, rubber and coffee, 27

Switzerland, elementary education, 167; college department, 172; provision for financing agricultural instruction in cantons, 172; extension service, 174; farm bookkeeping, 175; fruit canning, 176

Syria, extension work: Village Welfare Service, 86

Tanala of Madagascar, 12

Taylor, Carl C., 122

Tenancy in Pacific islands, 32

Tepoztlán, Mex., 120; work of Y.M.C.A., 130

Ting-Hsien Experiment, 50

Tingo Maria, Peru, 134

Tokelaus, 33

Transjordan, population, 78
Travancore State, India, Education Department, 72
Travel, peasants' attitude toward, 42
Tsou, P. W., 49, 50, 51
Tsou Ping Experiment, 51

United Kingdom, agricultural extension, 153-64; a local matter in England, 1; historical survey: agricultural education and research, 154; to outbreak of World War II, 155-58; County Agricultural Service and Agricultural Advisory Service, 155; number of farm holdings per county, 155; developments during World War II, 158-62; discussion groups: production records: milk testing, 160; plans for postwar developments, 162; extension worker, 163
United States, cooperation, 1; regional variations in economic rationalization of farmer, 149; well-to-do farmers control most of best land and economic resources, 150; agricultural extension, 180-92; population: number of families influenced: standards of living: racial origins, 180; Euro-American character of communities and organizations, 181; what is extension in U.S.? 181; objectives, 182; history and finance, 183; organization, 183 ff.; farmers' needs and extension methods, 185-91; work with organized groups, 188; ultimate purpose, 191
—— Agriculture, Department of: cooperation with State Agricultural Colleges, 183; farm demonstration work in South, 183
—— Farm Bureau, 150
—— Farm Security Administration, 150
—— Indian Service: program re Indian lands: extension service, 18
Usury in Pacific islands, 32

Veblen, Thornstein, 118
Venezuela, Indian population, 118
Veterinary service, in India, 66; in Latin America, 127

Victoria, agricultural extension, 5
Victory Garden movement in United Kingdom, 161
Village, cleavage between city and, 42; in China, 48; industries in India, 66; out of needs of, must grow reforms, 178

Wageningen, Neth., Agricultural College, 172
Wallis, Wilson D., quoted, 37
Wash-pot heaters, 189
"Week in the Garden, The," broadcast, 162
Weights and measurements of folk peoples and Euro-Americans, 139
Wells, H. G., 129
Western society, *see* Euro-American society
West Indies, Indian population, 118
Wilson, M. L., 48, 59
Winter Schools, 170, 171
Women, as extension workers, 2; condition in Balkans, 109; *see also* Farm women
Women's Institute, 158
World War II, disorganizing effects in Pacific islands, 19; developments during, in United Kingdom, 158
Wusi Teachers College, 51

Yen, Y. C. James, 50, 199
Y.M.C.A., National Committee of the Chinese, 49; Indian National Council, 66; Rural Demonstration Center in India, 75; work in area of Tepoztlán, 130
Young Farmers' Clubs, 157, 172
Youth, farm and home projects for, 3; Winter Schools: home-science schools, 171; 4-H club agents, 183, 184
Youth organizations, 4-H clubs, 4, 130; in Balkans, 112
Yugoslavia, land reform: population pressure, 101; national and religious minorities, 111
Yui, David, 49

Zurich, Technical College, Agricultural Department, 172
Zuni Indians, 14